SCR AND NEW TECHNOLOGY IN ELECTRIC RIG DRILLING: A SAFETY AND EFFICIENCY HANDBOOK

SCR AND NEW TECHNOLOGY IN ELECTRIC RIG DRILLING: A SAFETY AND EFFICIENCY HANDBOOK

Will L. McNair

PennWell Books
P.O. Box 21288, Tulsa, OK 74101

1421 South Sheridan/P.O. Box 1260
Tulsa, Oklahoma 74101

McNair, Will L.
SCR and new technology in electric rig drilling: a safety and efficiency handbook/Will L. McNair.
p. cm.
Includes index.
ISBN 0-87814-368-8
1. Oil well drilling rigs — Electric drilling. 2. Silicon-controlled rectifiers. I. Title.
TN871.2.M365 1991
622'.3381—dc20 91-8190
CIP

Printed in the United States of America

1 2 3 4 5 95 94 93 92 91

DEDICATION

To

H. L. "DUKE" ZINKGRAF

A consultant to the drilling industry, and formerly vice president of Sedco-Forex and president of Baylor Company. He was instrumental in advancing the state-of-the-art in rig designs using SCR systems, electric brakes, dynamically positioning rig systems and other technology.

and

CURTIS KRUSE

Former Director of the Petroleum Extension Service (PETEX) of the University of Texas, now retired. He provided training programs and publications that allowed drilling and production knowledge to be advanced throughout the petroleum industry.

CONTENTS

PREFACE

The use of electrical power in land and drilling rig operations has matured. Since its early use in the 1970s, the technology and use of electrical equipment have proven its reliability in one of the severest applications that can be envisioned.

An area that has been accepted almost unconditionally in the drilling industry is the silicon-controlled rectifier (SCR) variable speed drive. This system uses solid-state technology that provides power conversion and control of AC power to DC variable voltage power. The variable DC allows DC motors to operate as variable speed units to power drawworks, mud pumps, rotary tables, top drives, thrusters, cement pumps, winches, propulsion units, and others.

The first SCR system was designed and manufactured by the Baylor Company of Houston in 1966, and installed on the Offshore Company's offshore rig "Jubilee." Most modern-day rigs use AC-generated power and SCRs for variable speed drives. Not only are these systems reliable and efficient, they provide smooth, easily maintained control of speed, torque, and power to drilling functions.

With the downturn in the industry during the 1980s, many drilling rigs have been "stacked" and unused for many years. Even during this down period, the industry continued to investigate ways and methods to improve drilling performance, improve safety, and reduce costs. Most significantly, the top drive became a dominant product because of its ability to solve operational and safety problems such as reducing the possibility of a stuck bit by allowing continuous rotation of the drill string while circulating and pulling off bottom. Other new products came out of this period, including automated drilling rigs, automated breakout/makeup equipment, automatic pipehandling equipment, and others that removed personnel from the rig floor for safer operation.

In the early 1990s, horizontal drilling came into play that allowed the contractor to drill downward vertically before making a right-angle turn into multiple fractures and pay zones. The result was a significant improvement in oil and gas production from a single drilling program.

I see a significant new era beginning in drilling because of new technology that has been produced and proven in other industries. These new products include the personal computer, microprocessors, fiber optics, AC variable-frequency drives, and others. Among the applications that will come about from the use of these products are fully automated drilling rigs using microprocessors and electrical actuators, AC motors replacing DC motors in variable-speed applications, fiber optics replacing control cables in electrically noisy environments, and new functions replacing BOP controls, mud systems, and drawworks.

As an electrical engineer with a desire to provide new technology and teach others about it, I have often wondered about being in this business dominated by petroleum and mechanical engineers, high-risk investors, innovative drilling contractors, and producers who provide the primary funding for these endeavors. However, this may be the proper time to make a significant contribution to this industry through the application of electrical, electronic, and computer technology.

I especially want to thank my family for their support and encouragement in doing one of the things I enjoy most, and that is writing. This includes Sue; Mike, Carmen, Jennifer, Jeffrey, Jeremy; Bryan, Stacey, Christopher; Dianne, Anthony and Joshua. As I continue writing, I am sure the list will grow longer.

1 INTRODUCTION

Acceptance of electric power on drilling rigs has reached its maturity because of the significant applications resulting from its use. Applications using diesel/AC generators to provide three-phase and single-phase power throughout the rig has allowed many electrical loads to be added to the rig as compared to rig systems 15 to 20 years ago.

Among these loads has been the extensive use of electrical lighting of several types on the mast and derrick, in the critical mud tank areas, in quarters, and throughout the rig area. Constant-speed electrical motors are used for mixing, pumping, blowing, hoisting, and actuating in many areas on the rig. SCR units, the major users of electrical power, dominate most modern-day rigs where variable-speed power is needed. And most recently, automatic controls, instrumentation, and computers operating from the electrical power source are being applied in drilling operations.

Equipment using electrical power to function has resulted in significant improvements in the operation of the drilling rig. The concept of using multiple diesel engine/generator sets connected to a common AC electrical bus has reduced downtime significantly, since failure of one set will not shut down the power system. Using multiple SCR units that are selectable to the variable-speed DC motor loads also reduces the potential of downtime in the event a particular SCR unit fails. And the improvements of new controls using electrical equipment, including the top drive, downhole instrumentation, automatic pipehandling equipment, electrical pipe torque wrenches, and many others, show that electrical power applications offer significant benefits to the driller.

Knowing how to properly size, select, and apply electrical equipment on the rig becomes a serious consideration when we consider cost, size, weight, and function. Also, being familiar with the systems on the rig is part of the process of operating efficiently and economically to keep the cost of drilling to a minimum. Maintenance and repair costs can be held to a minimum, and rig downtime reduced, if we become familiar with the characteristics of the electrical equipment.

It is the purpose of this book to provide the necessary information to familiarize those involved in all aspects of drilling and rig operations with the electrical equipment used. From this information, improvement in operating and maintenance costs can be realized, allowing both the drilling contractor and producer/operator to benefit.

Functions of Land Drilling Rigs

By definition, a land drilling rig is generally an oil and gas structure/machine used on land, islands, platforms, or other surfaces where flotation or propulsion are not required for drilling or rig moves. These rigs have relatively simple equipment configurations, cost less and move easily from one site to another (Fig. 1–1).

The purpose of all drilling rigs is to create a near-cylindrical hole to a projected depth for producing oil or natural gas. To achieve this goal and to overcome a multitude of geological problems on the way to the desired depth requires substantial equipment and drilling skill. Drilling can truthfully be said to be a process of science, experience, and art.

The functions required of a rotary drilling rig usually include the following:

1. Turning the drillpipe and consequently, the bit, is accomplished by a rotating assembly. The *rotary table* using a kelly and the *top drive* suspended above the drillstring are the two major systems that perform this function. A rig may contain either one or both of these systems.
2. Hoisting the drillpipe out of the hole or lowering it into the hole and maintaining its position to control weight on the drill bit is accomplished with the *drawworks* system.
3. Providing the pressure and flow rate of the chemical "mud" is the function of the *mud pump*.

Fig. 1–1. Typical land drilling rig

4. Other functions on the rig can be said to support the above major drilling equipment and are referred to as *auxiliary* functions.

Major Rig Components

The major components used on land drilling rigs as noted above are the rotary table, top drive, drawworks, and mud pumps.

The Rotary Table

The principal mechanical rotating force, or torque, on the drillpipe is achieved through the rotary table located on the drill floor. It rotates on a

Fig. 1–2. Typical rotary table (Courtesy National Oilwell)

bearing surface and drives the "kelly" through bushings, and subsequently, the drillstring and drill bit. It also supports the slips when drillpipe is run or removed from the hole, and "spins out" tool joints when removing pipe.

The power for the rotary table is obtained directly from an independent electric drive motor or from a sprocket and chain arrangement coupled to the drawworks. Power for the rotary table is typically lower than that required for the drawworks and mud pumps and usually ranges from 400 HP to 1000 HP. The rotary table must be capable of operating in both directions of rotation, at variable speed, and be limited in its torque to avoid overtorquing and twisting off the drillpipe (Fig. 1–2).

The Top Drive

First introduced in the mid-1980s, the top drive has made significant inroads in supplanting the rotary table during drilling operations. The top drive consists of a DC motor operating as a variable-speed, variable-torque power source directly or through a gear transmission. Its speed is controlled by the driller from his console and eliminates the use of the kelly.

The primary advantage of the top drive is its ability to keep turning the drillstring and bit while circulating and pulling the bit off bottom. This action in itself has saved many stuck bit situations downhole and paid for itself even though they may cost from $500,000 to $1 million.

A disadvantage of some top drives is that the additional weight of the motor

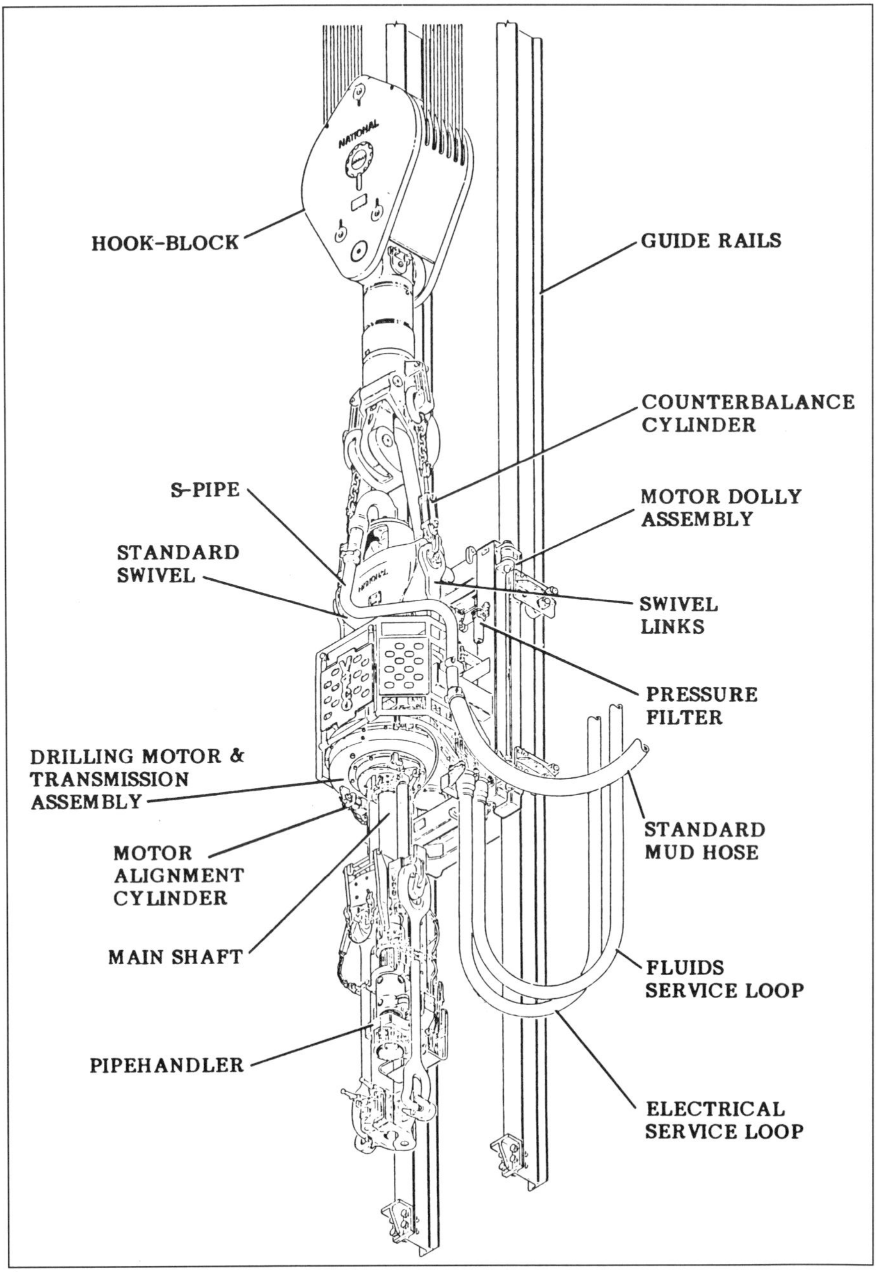

Fig. 1–3. A typical top drive (Courtesy of Varco International Inc.)

and transmission plus its size will result in modification of the derrick. For retrofits, such modification can be expensive, but top drives can be more easily accommodated on new rigs (Fig. 1–3).

The Drawworks

Handling drillpipe and drill collars and hoisting the casing for the hole is the primary function of the drawworks. It also raises and lowers the mast during rig-up or rig-down operations.

The drawworks consists of a large drum with Lebus grooving for properly wrapping and paying out the wire rope cable to the top of the mast. In this operation, the drawworks rotates in one direction to take up the cable and thus to raise the swivel and drillpipe components. By rotating in the opposite direction, the cable pays out to lower the swivel and drillpipe components.

The wire rope leaving the drawworks drum is referred to as the "fast line" because of its relatively high velocity during operation. Subsequent wire rope overlays in the mast pulleys to the traveling block result in a lower velocity but higher lifting force at the hook.

The drawworks also provides power for the catheads for making up and breaking out drillpipe and casing and provides power on some rigs for the rotary table through a chain and sprocket transmission. Friction clutches on the drum shaft, in addition to the drawworks transmission, optimize the speed and torque.

Electric motors coupled to the drive shaft provide mechanical power for the drawworks on electric rigs. Depending on the drilling depth and corresponding drawworks size, the system will have one, two, or even three 800–1200-HP drive motors. These motors are controlled in speed, direction, and torque with electrical controls connected to the motors by electrical cables. Since the drawworks is typically in the hazardous area of a drill rig, the motors must be either explosion proof or continuously purged with fresh air (Fig. 1–4).

Once the drawworks is up to speed with its load, slowing or powering down is accomplished with the electric brake directly coupled to the drawworks shaft (Fig. 1–4). Parking or emergency stopping is accomplished with a mechanical hand brake. The electric brake also aids in "drill assist" on some rigs by controlling the load descent on the drill bit during drilling. Since operation of the drawworks during drilling is at a relatively low speed, the electric brake can sometimes supply adequate braking torque at this speed to limit bit weight load. Before using the brake in this fashion, the brake manufacturer should be consulted to determine if improper operation or brake overheating can be avoided.

Fig. 1–4. Typical drawworks and cable assembly in rig (Courtesy Parker Drilling Co.)

Mud Pumps

Another major component in drilling operations is the mud circulating system. This system keeps the bore hole and drill bit clean, provides bit lubrication and cooling, and maintains adequate hydrostatic pressure on the formation to prevent hole cave-in or blowout due to abnormal bottomhole pressures (Fig. 1–5).

The equipment providing the mud flow is the mud pump. One or several mud pumps are used on a typical rig, depending on the desired circulating flow rate and mud weight. Primary power for these pumps on electric rigs is provided by electric motors in the 700–1200 HP range.

Depending on the mud pump size, each may have one or two motors driving the pump's mechanical input shaft. Variable speed of the electric motors is provided by electrical controls and is controlled from the throttle located on the driller's console. Variable flow rates can be achieved with this variable-speed system to compensate for downhole conditions and hole size. The pressure produced by the pumps is developed by the downhole pressure, weight of the mud fluid, and hole depth.

Mud Mixing Tanks

Initial mixing and agitation of the mud chemical occurs in the mud mixing

Fig. 1–5. Typical triplex mud pump (Courtesy National Oilwell)

tanks. Driving the mixing blades are constant-speed AC motors that blend the mud material and fluid to a homogeneous mixture of the desired weight and density prior to introduction into the circulating system. Other mud processing equipment on the tanks that use electrical power are the desanders, desilters, and shale shakers.

Auxiliary Motors and Lighting

Each rig varies in the number, size and type of auxiliary motors used. A typical system will have air compressors, supercharging pumps, water pumps, mud-mixing pumps, a degasser, cooling blowers, and mud pump rod oiler. These auxiliary motor functions typically use 5–100-HP AC motors, depending on the rig. In addition, lighting for engine area, rig grounds, tool pusher's quarters and mast are required for safety and operation. Additional electrical power for heaters, blowout preventer (BOP) controls, and other special functions is also required. Power for these auxiliary functions is supplied through a stepdown transformer to the motor control center where it is distributed to these loads.

Power Center

The heart of the electric land drilling rig is the power center, where control and distribution of the electric power occurs. Usually, the motor controls, switchgear, generator controls, motor control center, and lighting distribution are housed in an environmentally protected, skid-mounted enclosure or house. The house contains its own air cooling system, lighting, transformers, and plug board for external connections to the rig loads.

Depending on whether the system is a DC-DC Ward Leonard drive system or an AC-DC SCR system, the generator and variable-speed controls are housed in special cubicles in the power center. A typical land rig power center is shown in Figure 1–6 and a typical land rig SCR/generator control section is shown in Figure 1–7.

Fig. 1–6. Typical land rig power center (Courtesy General Electric Co.)

Fig. 1–7. Typical SCR/generator control section (Courtesy Baylor Co.)

Fig. 1–8. Typical diesel/generator set for land rigs (Courtesy Cummins Engine Co.)

Power Generation

The prime mover which converts petroleum fuels to mechanical energy is usually a diesel engine or gas turbine. Operating at essentially constant speed, these rotating machines produce the necessary torque to drive either DC generators or AC generators coupled to their shafts.

This engine-generator set is rigidly mounted to a skid structure and is protected against direct sun, snow, or rain. For land rigs that require portability, this is essential. Control of diesel speed with a governor and control of the generator voltage with a regulator is accomplished with controls located in the power center (Fig. 1–8).

Helicopter Rigs

A special variation in land rigs is the helicopter rig, identified because of its transportability and small packaging arrangement. It performs the same functions as the land rig but its weight is limited to a maximum so that a certain series of helicopter can lift and carry it to the drilling site. Increments of 5000, 6000, 7000, and 8000 pounds are typical of the size allocations that can be carried by medium to large helicopters. Sites not easily traveled to by conventional ground transportation (e.g., via truck or train) include forests, jungles, and other remote locations where oil and gas have been discovered. With the use of helicopters, these areas can be commercially developed for petroleum products (Fig. 1–9).

Functions of Offshore Drilling Rigs

While many of the drilling functions of offshore rigs are similar to land rigs, the offshore rig has several configurations and typically requires additional functions for its operations. Each type of rig is characterized by its application and operating features. A survey conducted in 1990 showed the following mix of worldwide offshore rigs: platforms, drill tenders, submersibles, drillships, barges, semisubmersibles and jackups. The jackup represents the most popular type in use because of its relatively low cost in variable water depths.

Submersibles

Used primarily in calm waters such as the Gulf of Mexico, these units are

Fig. 1–9. Helicopter TBA Rigs (Transportable By Anything) (Courtesy Parker Drilling Co.)

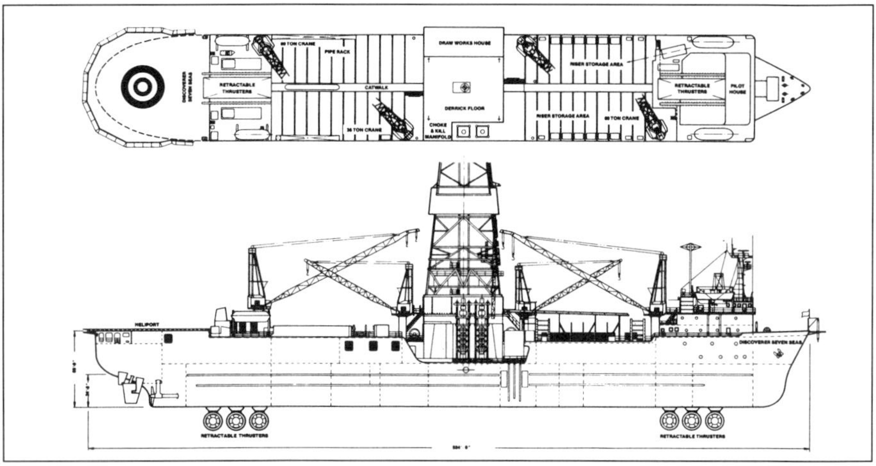

Fig. 1–10. Typical dynamically positioned drillship (Courtesy Offshore International S.A.)

typically placed in water depths of less than 100 ft, which limits their drilling assignments. Drilling depths range 18,000–30,000 ft, but the typical depth is 20,000 ft. Most of these rigs were built in the 1950s and generally use mechanical drives or Ward-Leonard DC-DC systems for drilling power. Diesel engines are typically the primary movers on these rigs, and drilling functions involve the rotary table, drawworks and mud pumps. Other equipment used on these rigs include cranes and mooring winches, some with electric drives. Location moves are accomplished by towing from one site to another, and sleeping and dining quarters are provided.

Drillships and Barges

Most of these rigs move from one site to another through self-propulsion, but some are towed. In addition, some are highly sophisticated vessels with dynamic positioning capabilities, heave compensators, and environmental instrumentation. Operation in water depths of 3300 ft or more and drilling depths of 30,000 ft can be achieved with units equipped with dynamic positioning. Many of these vessels, however, can drill in water depths of 600 ft with multiple-anchor mooring. These systems were constructed in the 1970s and use electric power for drilling, mooring, and thruster functions (Fig. 1–10).

Electrical equipment used on these rigs includes diesel-electric power

Fig. 1–11. Dynamically positioned drillship SEDCO-Forex 472 (Courtesy SEDCO-Forex)

generation; thrusters with electric motors; drawworks, rotary table, mud and cement pumps with electric drives; mooring winches with electric drives; and cranes with electric drives. A sophisticated, dynamically positioned mining ship is shown in Figure 1–11.

Semisubmersibles

Deepwater exploration and drilling are ideal applications for semisubmersibles since they can operate in 35–72-ft wave heights. The buoyancy chamber is placed below the water surface to minimize the effects of wave forces and to stabilize the drilling equipment. Cost of these vessels is high. Some semis combine self-propulsion with dynamic positioning for sophisticated drilling in water depths of up to 6000 ft and drill depths of up to 30,000 ft. Other equipment provided on these vessels includes drawworks, mud pumps, and rotary tables with electric drives; cranes with electric or diesel drives; winches with electric drives; thrusters with diesel or electric drives; and diesel-electric power generation. Figure 1–12 shows a dynamically positioned semisubmersible, and Figure 1–13 shows a typical submersible.

Fig. 1–12. Semisubmersible drill rig—dynamically positioned (Courtesy Marathon LeTourneau Co.)

Fig. 1–13. Semisubmersible drilling vessel (Courtesy Sedco-Forex)

Fig. 1–14. Typical jackup drilling rig (Courtesy Rowan International, Inc.)

Jackups

The jackup drilling rig combines economy, performance, and flexibility in medium-depth waters. It drills in water depths of 70–370 ft and to drilling depths of 30,000 ft. After being towed or self-propelled to a drilling site, a jacking arrangement lowers the support legs or columns to the seafloor. Using the seafloor as a foundation, the platform is then raised 15–50 ft above the water level with the jacking system. The jackup uses electric drives to power drawworks, mud pumps, rotary table, winches, cranes, and jacking units (Fig. 1–14).

An Evaluation of Electric-Powered Rigs

For each rig type there is a particular power system that best suits that rig. In most cases, the rig will be configured with a diesel-AC generator set for power generation and SCR AC-DC variable-speed controls for DC traction motors. However, it is useful to consider possible alternatives to choose the most appropriate power system for a particular rig.

The single most important question when considering various rig drives and power systems is, "What does this particular system offer?" Several points must be evaluated before a decision is made. The following checklist can help evaluate the various power systems available:

1. Define the operating parameters of the drill rig: i.e., drilling depth, drilling environment, type of rig.
2. Define the rotary table, top drive, drawworks, and mud pumps required and their associated drive motors.
3. Define the auxiliary motor functions, rig lighting, heaters, and other electrical equipment required by horsepower or kilowatt (kw) rating.
4. On offshore vessels, determine the thrust requirements, winch ratings, crane ratings, and jacking system ratings, if applicable.
5. Summarize all electric power loads from this equipment list. Adjust these ratings for the required accelerating power, equipment power factor, and use or duty cycle. This information determines the total kilovolt-amperes (kva) required by the AC generators and the subsequent kw and HP required by the diesel engines.
6. Considering load distribution, the need for redundancy, and diesel gen-

erator ratings and availability, select the electrical power generating equipment for the application.

7. Once the diesel generator set is selected, specify the engine actuator, governor, and generator regulating system.
8. Establish the rig layout and design the cable distribution system, rig lighting, local start-stop stations, and other equipment.

After this evaluation is made, a cost and performance evaluation of the rig should be made.

When considering cost, evaluate the following items to obtain an overall picture of the particular rig set-up:

- Initial equipment and rig component costs
- Cost to construct rig
- Costs to interface various manufacturers' equipment into an operating entity
- Rig-up and rig-down costs, including cost of moving
- Operating costs, including labor, fuel, expendable items, and training
- Replacement costs of fast-wearing items
- Maintenance and repair costs, including lost revenue during downtime
- Evaluate costs based on individual drilling jobs over the long term, including material, labor, and fuel cost increases

When considering performance, quantify those areas that significantly influence cost and schedule factors.

1. Determine how fast a rig-up/rig-down can be accomplished; evaluate cabling, mechanical equipment realignment, and moving capability.
2. Calculate the overall equipment operating efficiency as a function of hole depth. Computing 2500 ft increments helps the drilling contractor/owner determine fuel costs as drilling progresses. Consideration for horizontal as well as vertical drilling should be made. This analysis also aids the planning process once the rig is placed in operation.
3. Estimate the drilling speed of a particular system in the lease geology.
4. Estimate the downtime projected on a particular job for the equipment

under consideration. Include in the evaluation the time to repair and maintain particular components of the rig.

If a comparison of mechanical compound rigs and DC-DC rotating sets is made with AC generators and SCR equipment, the latter will obviously have many advantages. Using the performance and cost checklist, the systems will generally compare to one another as shown in Table 1–1.

Table 1–1
Comparison of Rig Power Systems

Function	Mechanical	DC/DC	AC-SCR
Power for rig lighting, AC motors, and auxiliaries	Requires separate AC generator	Requires separate AC gentor	Uses existing AC generator with transformer
Prime mover redundancy	Requires standby engine as backup	Requires standby engine as backup	Removing one AC gen will not not stop operation
Load assignment flexibility	Not available	Each DC motor can be assigned to DC generator	Flexible assignment capability
High maintenance items	Diesel engines, mechanical drive transmission chains, belts, alignment	Diesel engines, DC generators, DC motors	Diesel engines, DC motors
Flexibility in equipment location	Dedicated locations for coupling to compound shaft	Unrestricted flexibility	Unrestricted flexibility
Diesel engine utilization	Inefficient use at light loads in compounds	Inefficient use of dedicated engine at light DC generator loads	Efficient use of time shared engines
Flexibility in diesel engine selection	Must be matched to maximum compound load	Must be matched to maximum DC generator load	Wide selection of engine-generator arrangements

Function	Mechanical	DC/DC	AC-SCR
Power for AC motors	Must have dedicated AC generator set	Must have dedicated AC generator set	Can obtain AC from system bus
Operation from AC utility supply	Not available	Not available	Easily adaptable
Engine shock loading	High	Can be controlled and is usually low	Can be controlled and is low
Average efficiency from engine shaft to drilling machinery input	75%	88%	87%
Relative rig first cost	1.1	1.1	1.0
Repair and maintenance training required	Mechanic knowledge	Electrician knowledge	Electrical and some electronic knowledge
Rig moves of land rigs	Requires mechanical alignment	Requires minimal attention	Requires minimal attention
Motor speed range	Limited	0 to full speed	0 to full speed
Torque and speed control and limits	Not easily controlled	Controllable	Controllable
Diesel engine location relative to wellbore	Within 40 ft of wellbore	150–250 ft of wellbore	150–250 ft of wellbore
Rig floor noise level	High with diesel engine noise	Low with drilling equipment noise only	Low with drilling equipment noise only

The evaluation of Table 1–1 can be extended to other areas by obtaining data from the equipment manufacturers relative to performance and cost. Many suppliers of high-quality electrical equipment for drilling exist in the United States as well as in other countries. Electric generators, motor control centers as well as SCR systems can be obtained from several American manufacturers, European manufacturers and Far East manufacturers.The electric rig may be slightly higher in first cost than other types, but considerable fuel and maintenance cost savings can be realized during operation. Additionally, improved drilling efficiency and operation flexibility can be obtained from the AC/SCR system. An illustration of the economics involved in the purchase cost is shown in the following example of a 12,000 ft land rig.

Table 1–2
Cost Comparison of Mechanical vs SCR Electric Rig (%)

	Electric	**Mechanical**
Drawworks and accessories	7.7	10.5
Rig power controls or compound	16.7	10.5
Mast and substructure	21.1	21.1
Mud pumps and accessories	11.7	10.4
Mud system equipment	5.5	5.5
Handling tools	1.0	1.0
Rotary and traveling equipment	5.1	5.1
Instrumentation	1.0	1.0
Drillstring (12,000 ft)	12.1	12.1
Rig air system	0.5	0.5
Miscellaneous rig equipment	4.5	4.5
BOP	7.7	7.7
Shop and hand tools	0.2	0.2
Rig-up/rig-down	7.2	6.2
Electrical power and distribution	0.0	2.7
TOTAL (% of Mechanical Costs)	102.0	100.0

Summary of Comparisons

When a comparison of initial costs and operating costs are made between the electric SCR rig and the mechanical rig, several interesting facts emerge. A summary of a typical analysis for a 12,000-ft land rig is given below:

Initial cost — Electric rig initially costs approximately 2% more than the mechanical rig.

Fuel savings — Up to 25% fuel savings of electric rig over mechanical rig.

Savings in lost revenue due to improved reliability — Approximate 5% improvement by electric rig.

Savings in maintenance cost — Approximate 5% improvement by electric rig.

Annual savings — Approximately 15% overall operating cost savings per year by electric rig.

Bibliography

Harris, L.M. *An Introduction to Deepwater Floating Drilling Operations,* Tulsa, Petroleum Publishing Co., 1972.

Neel, Marvin M., et al. "Application of Thyristor Control for a Drilling Rig." Paper presented to the American Association of Oilwell Drilling Contractors in Dallas, September 1970.

Stone, W. M. "A Systems Approach to Electric Land Rigs." Paper presented to IADC Drilling Technology Conference in Denver, March 1979.

Webb, Glenn W. "A Comparison of Mechanical and Electrical Drives for Land Drilling Rigs." Paper presented to 1977 Drilling Technology Conference, Houston, 1977.

2 ELECTRIC POWER GENERATION AND CONTROL

Unless a drilling rig is in an area where electrical utility power is relatively inexpensive and directly available to deliver 1–3 megawatts (mw) of power, the rig must provide its own source of power. The first source of this generated power must be obtained from a fuel that converts its energy into mechanical speed and torque. The steam engine was one of the first methods used for the prime mover. Because of the inherent problems associated with steam, the system eventually turned to the diesel engine and gas turbines as the most popular prime movers.

To produce electrical power, the prime mover is coupled with an electrical generator to provide the mechanical input power. When excited from a separate electrical source, these generators produce electrical power in the form of DC or AC.

Prime Movers

Gas Turbines

Gas turbines are prime movers characterized by their high speed (up to 42,000 rpm), small size and weight, and high noise level. While several of these units have been used in offshore drilling operations, few have been used on land.

Their disadvantages exhibit themselves because of the speed mismatch between DC generators and AC alternators since they typically operate at 900–1800 rpm. Speeds above 2500 rpm are prohibitive since they can cause damage

Fig. 2–1 Gas turbine-AC generator system (Courtesy Cummins Engine Co.)

due to winding separation, brush bounce and other problems. Conversely, the turbines operate at very high speeds thereby requiring speed reduction gearing for compatibility.

A gear train reducing the turbine speed from 18,000–42,000 rpm to the required 1200 rpm, for example, would be costly, inefficient, and would add weight to the assembly. Therefore, careful design consideration must be made before deciding to use turbine-generator sets (Fig. 2–1).

Diesel Engines

Diesel engines are characterized by their low speed, limited speed range, relatively low maintenance, and general availability. Selection of diesel engines to drive electric generators is an obvious choice because their similar operating speeds allow direct coupling. The torque and horsepower of both are compatible, and control of engine and generator speed allows relatively easy control of generator ouput power.

Operation of a diesel engine, such as that shown in Figure 2–2, differs from a gasoline engine; no spark plug is required to achieve combustion of the fuel. A gasoline engine injects the gasoline and air mixture into the combustion chamber, then rapidly heats to a combustible temperature with the electrical spark of a spark plug. After the small portion of gas-air mixture around the spark plug starts to burn, it instantly heats and ignites the remaining mixture.

The diesel engine operates on the principle that compressing any gas will cause it to heat to a very high temperature. More importantly, the diesel engine starts the first stroke of its cycle by allowing only air to enter the engine cylinder at a pressure of around 500 pounds per square inch (psi). This pressure heats the air to 1000° F during this compression stroke and shortly afterward mixes with the diesel oil spray to start its combustion. Power is then developed during the power stroke, and burnt gases are exhausted at the end of the cycle.

The power developed by a diesel engine is a function of operating altitude, since a definite weight of air must be supplied to burn a pound of fuel. The higher the altitude, the lighter the air density; thus, a lower horsepower is developed. Most engines are rated based on S.A.E. J816b test conditions at 500 ft, 29 in. Hg, 85°F. But operation at high altitudes, e.g., the Rocky Mountains, reduces horsepower. Refer to the manufacturer to determine this rating or for modifications such as supercharging to improve ratings at high altitudes.

The operating efficiency of a diesel engine varies with air density. It is also a function of its load relative to its rating. Figure 2–3 is a typical curve of a diesel engine showing fuel economy vs load. It shows that an engine can produce 13.3 kilowatt-hour (kw-h) for every gallon of fuel oil when operating at 100% of its rating but produces only 11 kw-h per gal at 50% of its rating. Thus, running diesels near their rating is the most efficient operating method.

Starting a diesel engine requires that it reach a speed sufficient to bring the compressed air charge to the oil- igniting temperature. To achieve this speed,

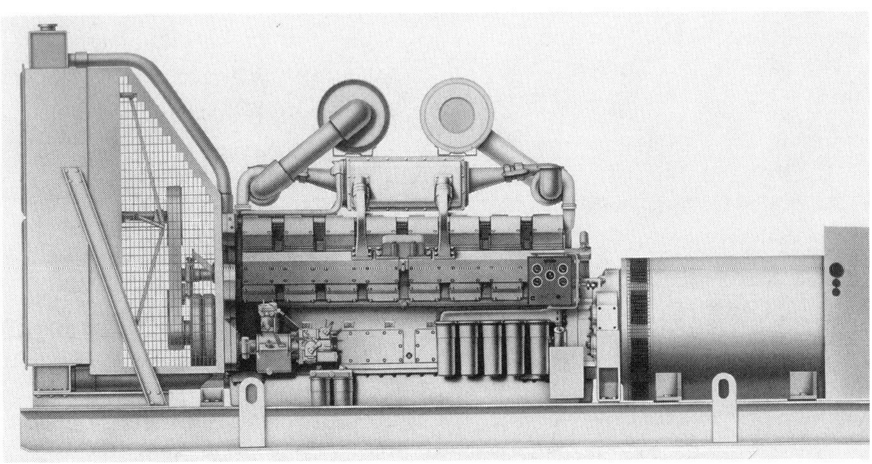

Fig. 2–2 Typical diesel engine with AC Generator (Courtesy of Caterpillar Co.)

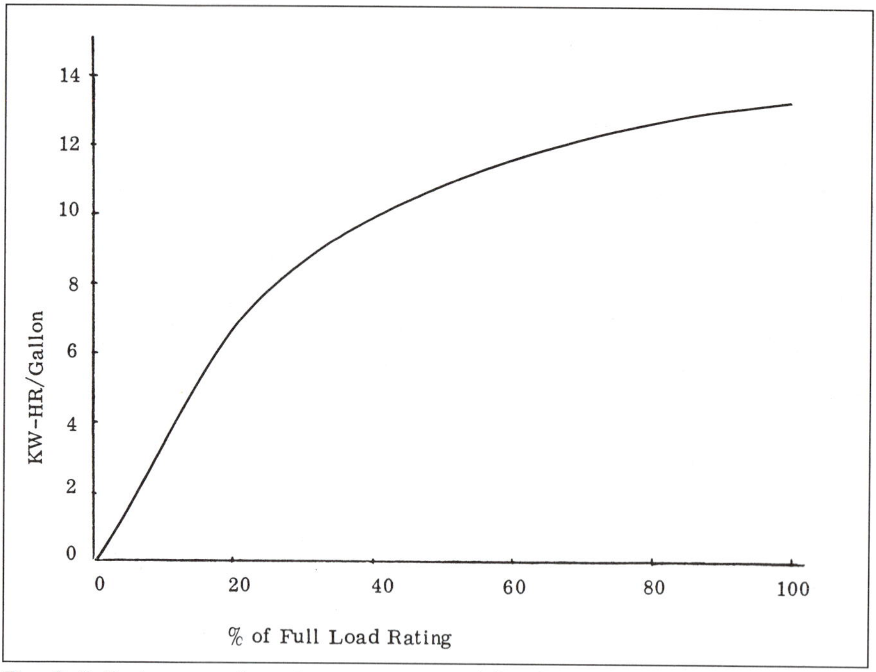

Fig. 2–3 Fuel consumption rate versus percent of diesel engine rating at sea level

there are two possible methods to use: air-starting by injecting air into the diesel cylinders from a separate drive air compressor and air storage tanks at 250–300 psi, and auxiliary motor starting from an electric DC motor, a small gas engine, or pneumatic motor.

The higher the cranking speed achieved, the higher the compression and the faster the start of the diesel engine will be as shown in Figure 2–4.

Once the engine is cranked, the engine speed must be regulated. On electric drilling rigs employing AC generators, this speed must be closely held for proper frequency control as well as for load sharing between paralleled machines. Most engines incorporate both a mechanical centrifugal ball governor as a backup and an electronic governor for primary speed control.

Since a diesel engine is primarily a constant-torque machine for any one fuel control setting, adjusted fuel flow is required to maintain constant speed as load changes. The speed governor monitors the engine speed and automatically

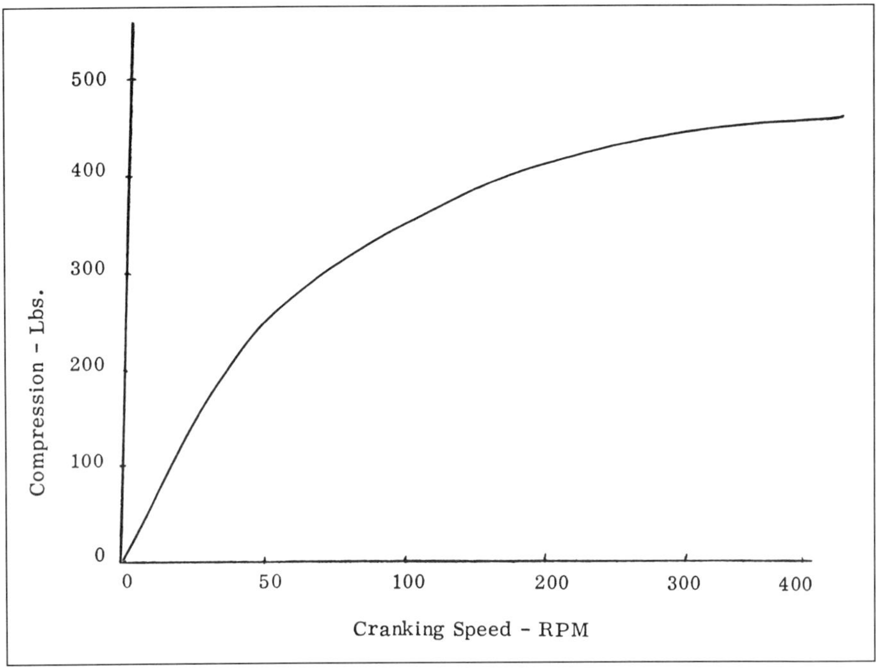

Fig. 2–4 Diesel engine cranking speed vs engine compression

controls the amount of fuel injected into the engine cylinder. As the load increases, the speed momentarily decreases, creating a speed "error" in the governor. This error causes the fuel rack to adjust for more engine fuel and return to its original speed. The reverse occurs for load decreases.

Governors and their associated fuel actuators for engines used on drilling rigs are usually supplied with the engines or the generator/engine switchgear supplier. The governors must be compatible with the actuator on the engine for proper operation to occur. Most governors are electronic, with a magnetic pickup and toothed wheel to provide the speed feedback control.

In the constant-speed mode required for AC generators, speed control operation is called "isochronous."

The desired speed of the engine is set with a control adjustment by the operator, establishing the speed "reference." Electric governor output automatically adjusts the proportional actuator to set the desired fuel flow. The actual speed is sensed by the magnetic pickup at the engine, and this signal is fed back to the governor for comparison with the reference. Once the speed reference and feedback are satisfied with the minimized speed error, the speed remains relatively constant.

Other control functions may be added to speed sensing when coupled to an AC generator: frequency sensing, load sensing, and acceleration rate control. The electric governor receives all of these input signals from the sensors and establishes control of the engine based on the priority of these signal demands.

Some electric governors provide proportional signals to the fuel actuator of the same polarity, while other governors provide a positive or negative signal polarity to adjust the actuator. Either method is satisfactory, but the governor and actuator must be compatible. Should the electric governor fail, the mechanical flyball governor should be set as an overspeed device to protect against engine runaway.

Cooling a diesel engine on drill rigs is usually accomplished with a heat-transfer system using water flowing through the engine jackets to a fan-cooled radiator. Often, the engine manufacturer supplies the cooling system with the engine along with coolant recommendations. For long life and low maintenance, the inlet-outlet cooling water temperature differential should be kept as low as possible. The lower the temperature rise, the lower the mechanical stresses in the engine will be. A general rule of thumb is that temperature differential should be 12°F–20°F with an outlet temperature not over 180°F and an inlet temperature not lower than 60°F. The manufacturer's unit information provides a more precise figure to be followed.

The quality of the cooling water or fluid minimizes engine jacket scale and corrosion. Closed cooling systems should employ pure water with additives to reduce these problems and maintain good heat transfer.

The useful life of a diesel engine averages 12 years, depending on the operating care and maintenance it receives. However, operating life of 30 years is possible if proper lubrication, cooling, and repairs are maintained. Other considerations, such as depreciating the original investment over a certain period or replacing engines with improved units, may in actuality determine the useful life of an engine.

The mechanical power delivered to the shaft of the engine is the final result of the heat energy to mechanical energy conversion. Power available and obtained from the shaft is expressed in terms of horsepower or kilowatts and is related by the following expressions:

$$1\ \text{HP} = 746\ \text{watts} = 0.746\ \text{kilowatts}$$

$$\text{HP} = \frac{(\text{torque})(\text{speed})}{5252}$$

where torque at shaft, ft-lb speed at shaft, rpm

DC Generators

A System View

The primary means of controlling DC drilling and thruster motors during the 1950s and 60s was power developed by DC generators. Coupled directly to the shaft of a diesel engine, the DC generator would produce variable DC power to control the speed and torque of the dedicated drilling motor. Many of these systems are still being used.

If a DC shunt and DC generator of similar rating, frame size, and manufacturer were placed side by side and observed for differences, there would be few discrepancies. In fact, except for slight design differences in the commutator and armature winding, they are identical. The way the armature winding, field winding, and shaft accept or deliver power distinguishes them in operation.

A DC shunt receives a constant field voltage (or field excitation) and develops speed and torque at its shaft based on the amount of armature voltage. It converts electrical power to mechanical power. A DC shunt field generator, on the other hand, receives a constant shaft speed from a prime mover and develops DC armature voltage and current at its terminals based on the amount of applied field voltage. It converts mechanical power to electrical power.

The general equation that relates generated DC voltage at the armature is:

$$E = K\, I_{FG}\, N_{G}, \text{ volts DC}$$

where
E = generated armature voltage
K = machine design constant
I_{FG} = shunt field current, amperes
N_G = generator shaft speed, rpm

At a constant speed, the generated voltage is directly proportional to the field current. This principle is the one on which the DC/DC speed-control system of Figure 2–5 is based and is generally known as the Ward-Leonard Speed-Control System after the individuals who first conceived this concept.

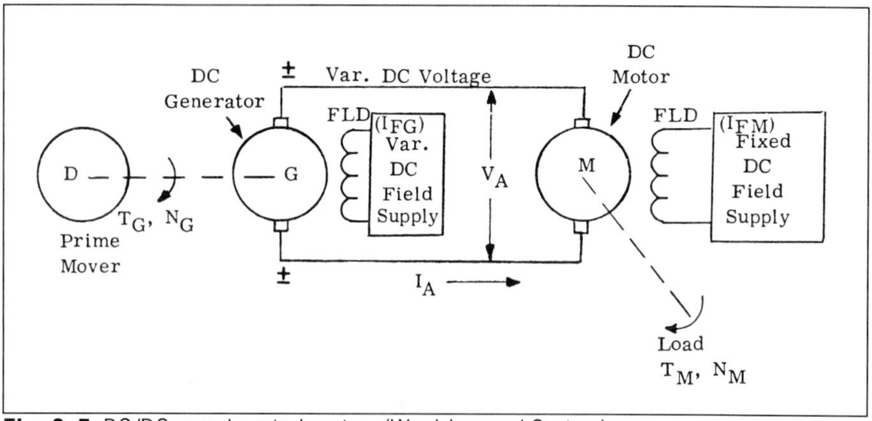

Fig. 2–5 DC/DC speed control system (Ward-Leonard System)

Four equations approximate the various equipment interfaces in the DC/DC system.

1. Torque load imposed on the diesel engine shaft by the DC generator:

$$T_G = K_G I_{FG} I_A, \text{ ft-lb}$$

where
K_G = DC generator machine constant
I_{FG} = DC generator field current, amperes
I_A = DC generator/DC motor armature current, amp

2. Armature voltage applied to DC motor terminals:

$$V_A = E - I_A R_{AG} - V_{BG}, \text{ volts DC}$$

where
E = generated DC voltage
R_{AG} = DC generator armature resistance, ohms
V_{BG} = DC generator brush drop, volts

3. Speed of DC motor:

$$N_M = \frac{V_A - I_A R_{AM} - V_{BM}}{I_{FM}} K_M, \text{ rpm}$$

where
R_{AM} = DC motor armature resistance, ohms
V_{BM} = DC motor brush drop, volts
I_{FM} = DC motor field current, amperes
K_M = motor constant

4. Torque developed by DC motor:

$$T_M = K_M I_{FM} I_A, \text{ ft-lb}$$

where K_M = DC motor constant

Speed control of the drilling machinery is accomplished in the DC/DC system by varying the field current to the DC generator. The precision of speed control can be improved by sensing the output DC motor shaft with a tachometer and feeding this data back to the generator field supply for any necessary correction or adjustment to maintain constant speed.

The DC Machine

A DC generator produces DC electrical power from mechanical power (Fig. 2–6). This is achieved by an electrical conductor passing through a magnetic

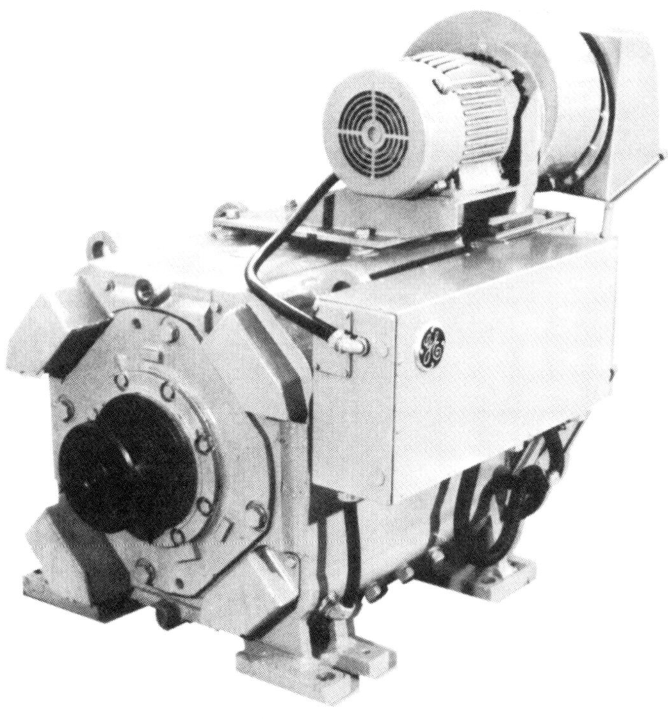

Fig. 2–6 Typical DC generator (Courtesy General Electric Co.)

field which develops a voltage in the conductor. The magnitude of the voltage induced in the conductor is a function of the magnetic field intensity and its rate of change. Expressed analytically, the induced voltage is:

$$E = K\, d\O/dt, \text{ volts}$$

where K = constant consisting of the number of turns and conversion factor

$d\O/dT$ = rate of change of magnetic flux

Relating this principle to a DC generator, the magnetic field is created by the field coils of the stators. Exciting these field coils with DC current produces a magnetic field around the periphery of the machine.

The rotor or armature consists of heavy conductors and relatively few turns.

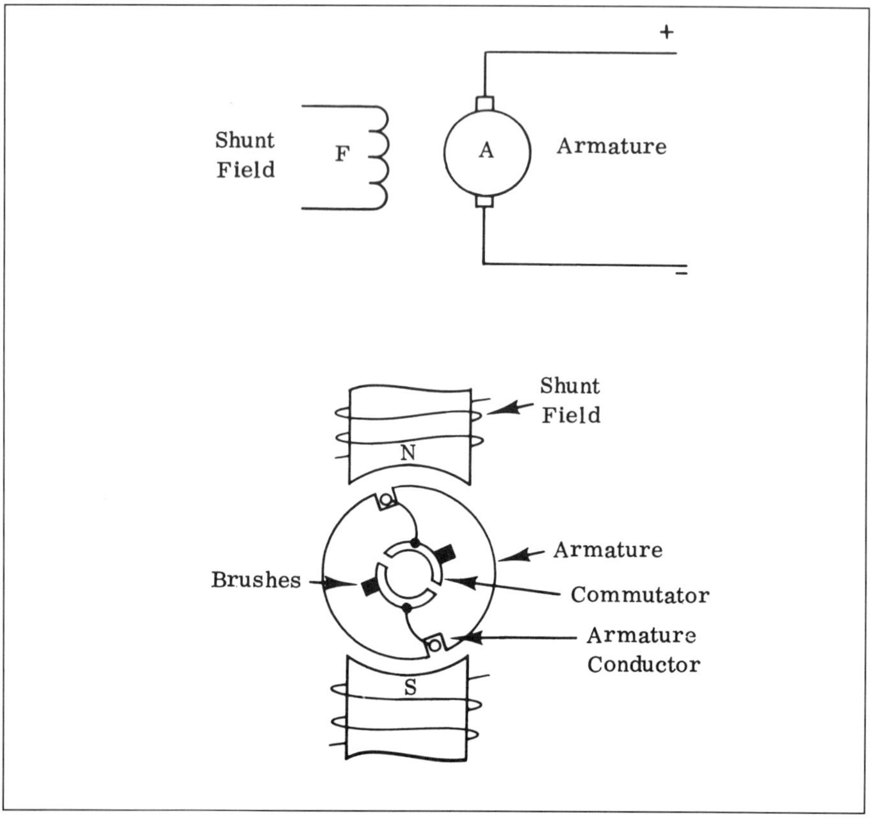

Fig. 2-7 Separately excited, shunt-wound DC generator

As the prime mover rotates the shaft and armature, the armature conductors enter and leave the magnetic field established by the field coils. This change in the rate of flux linking with each armature conductor produces a voltage in the conductor. The faster the armature rotates, the greater the voltage will be.

The segmented commutator and brushes conduct the generated power from the rotating armature to external loads. Each end of an armature conductor is connected to a segment of the commutator, which is a group of insulated copper bars on one end of the shaft. Brushes of carbon graphite composition slide in contact with the commutator and conduct the power from the armature (Fig. 2–6).

As the brush overlaps one segment to the other, the voltage is low and little current flows between segments. As the coil rotates, the positive brush always receives current from the coil half passing through the positive voltage-induced area of the field. Therefore, the brush and commutator act as a rectifier to keep the voltage at one brush positive and the other brush negative.

The friction between the brush and commutator must be kept low enough to minimize brush wear and firm enough to prevent bouncing and creating arcing and pitting on the commutator surface. The higher the current, the larger the brush assembly and commutator must be to maintain the proper current density in these components.

Most generating machines used in the drilling industry are separately excited, shunt-wound types (Fig. 2–7) as compared to other configurations referred to as self-excited, shunt-wound types or self-excited, series-wound types.

AC Generators

Voltage Generation

The conversion of mechanical energy from the prime mover to a generator producing alternating, three-phase voltage is important since most domestic electrical equipment is designed for AC power.

An AC generator or alternator is similar to a DC generator; it requires DC field current for its excitation. It also has an armature with multiple turns of wire to generate the electrical voltage as the rotor is turned by the prime mover.

Beyond this point, differences emerge in the comparison. The brushes and commutator of the DC machine are replaced with slip rings which eliminate the commutator and brush wear problems. Since the field current is only 2%–3% of the armature current, the armature winding on large machines are placed on the stationary part of the machine and allow the slip rings to handle the rotating field or the brushless field exciters on the rotor.

Generating and distributing three-phase power is more efficient using steel and copper than single-phase generation. In fact, most AC rig motors and other equipment are three-phase.

A simple three-phase, two-pole generator has three sets or groups of armature coils spaced 120° apart with each group brought out to separate connection points. The generated voltages in each group have identical wave shapes but vary in time phase with respect to each other since they enter the field flux at different times (Fig. 2–8).

A two-pole generator produces a positive and a negative half-cycle for each phase of each revolution of the armature. In this case, the mechanical degrees are the same as the electrical degrees; i.e., one shaft revolution equals 360°, which is equal to a full voltage sine wave of 360°. The phase difference between each of the three phases is 120 electrical and mechanical degrees.

A four-pole generator produces two full sine waves per phase for each revolution of the armature, and a six-pole generator produces three full sine waves per phase for each revolution. The relationship between the number of machine poles, shaft speed, and the frequency of the generated voltage is expressed as follows:

$$f = \frac{PN}{120}, \text{ cycles/second (Hertz or Hz)}$$

where f = frequency of voltage, Hz
P = number of machine poles
N = shaft speed, rpm

To produce a frequency of 60 Hz, a two-pole machine must operate at 3600 rpm, a four-pole machine must operate at 1800 rpm, and a six-pole machine must operate at 1200 rpm.

There are voltage transmission advantages in connecting the three generator phases in "delta" or "wye"/"star" configurations instead of bringing each phase out separately. Using each phase separately results in bringing six wires out (Fig. 2–9), while wire savings result with the delta or wye connection.

A delta-connected generator allows only three connections to be run for load connections, saving one connection over a wye system. However, the system does not have a neutral point where a ground can be established to provide line-line and line-neutral voltages. Although a delta-connected generator has its three windings connected in series, no circulating current flows

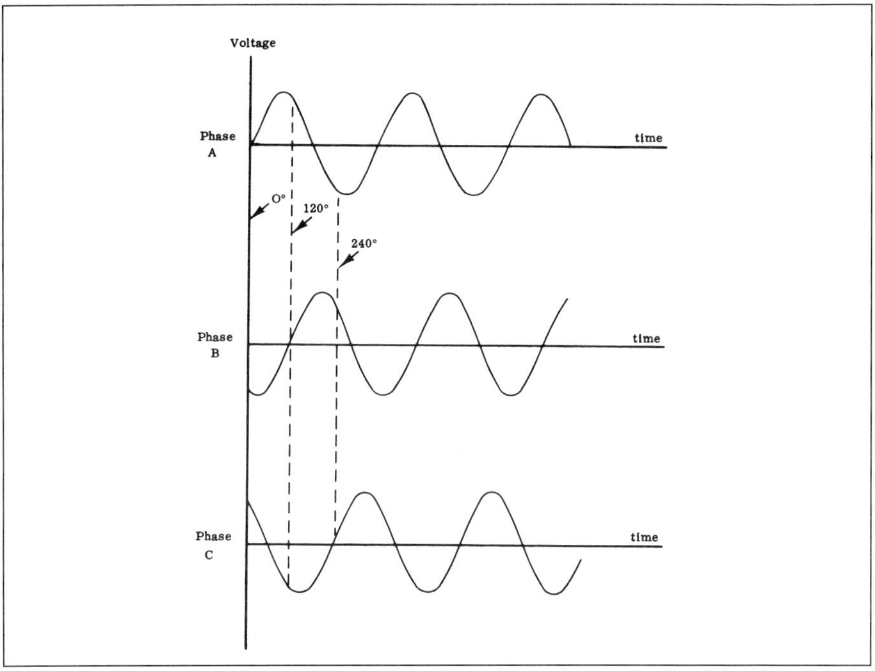

Fig. 2-8 Three-phase generated voltages

between the windings since the sum of the three-phase voltages is zero. In other words, no voltage difference exists to develop a winding circulating current.

Voltage and current relationships of the delta connection are:

V_L = Vø, volts

where VL = line-line voltages = $V_ø$ = phase voltages

I_L = $\sqrt{3}$ Iø, amperes

where I_L = line current = 2 cos 30° x phase current

= $\sqrt{3}$ phase current

because of the 120° phase displacement.

A wye connection joins one end of each generator phase coil at a common or neutral point. The voltage of each coil or phase is vectorally added to calculate the line-line voltage. Grounding the neutral in some applications reduces high voltage exposure to personnel.

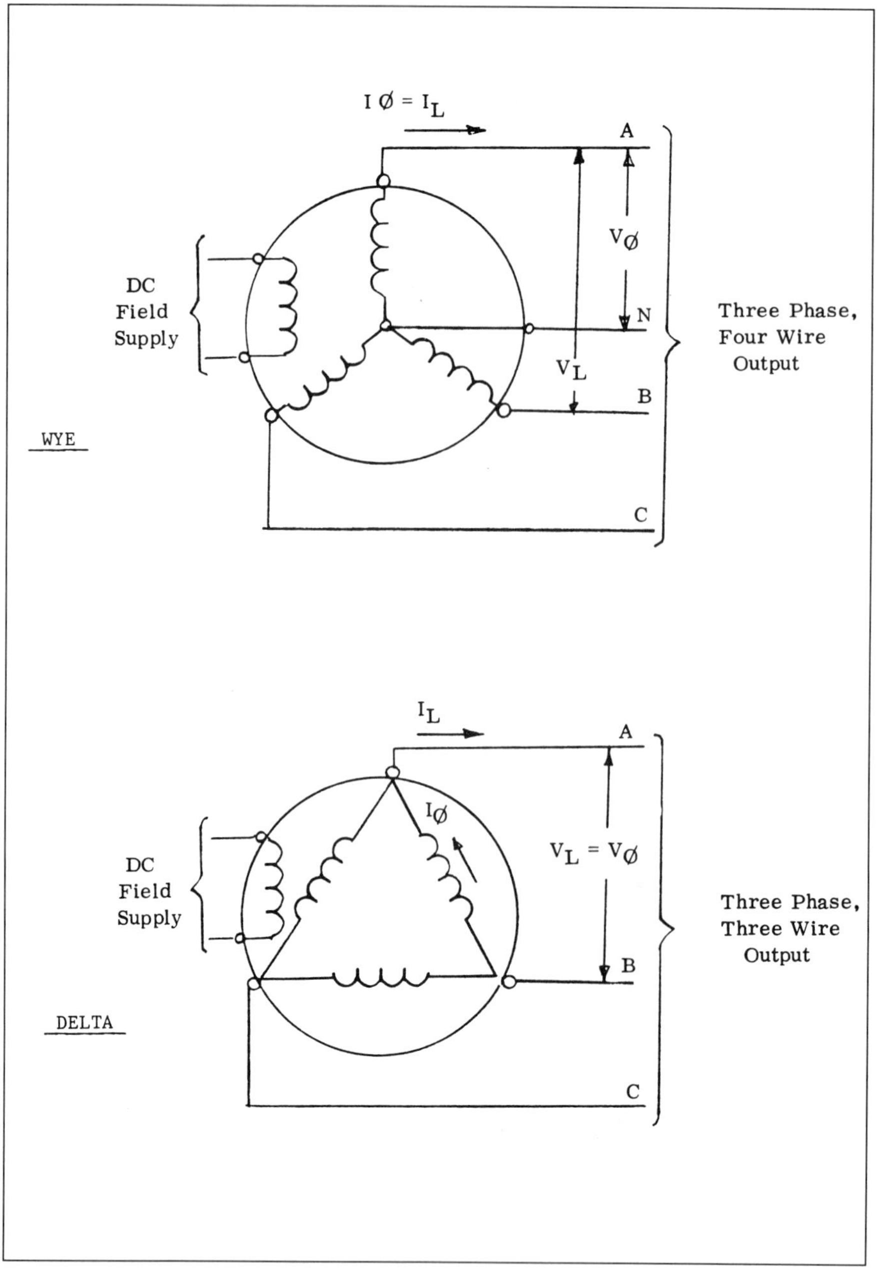

Fig. 2–9 Delta and wye generator connections

The voltage and current relationships for the wye connection are:

$$V_L = \sqrt{3}\, V\phi, \text{ volts}$$

where V_L = line voltage = 2 cos 30° x phase voltage
= $\sqrt{3}$ phase voltage,
because of 120° phase displacement.

$$I_L = I\phi, \text{ amperes}$$

where I_L = line current = phase current

An example of wye generator connection is a system producing 208/120 volts. The line-line voltages are 208 volts each and the line-to-neutral or phase voltages are 120 volts. The system allows operation of three-phase, 208-volt AC motors and distributes 120 volts power to lighting and other single-phase loads. By maintaining a balance of single-phase loads on each of the three phases, the generator can produce its full rating economically without overheating.

Voltage Regulation

Voltage regulation maintains the generated voltage at a relatively constant value. Slight speed variations, load changes, and field temperature changes influence the output voltage. Sensing the output AC voltage and comparing this to a reference in a closed loop controller allows a means of controlling the voltage. If an error exists between the reference and feedback, the controller can adjust the generator field current to correct the voltage. Voltage regulators can also sense output current for protecting and load sharing between generators in parallel.

Kva vs Kw

The size and rating of an AC generator are based on the generator's kva capacity. Kva is expressed analytically for three-phase generators as:

$$\text{kva} = \frac{V_L I_L \sqrt{3}}{1000} \quad \text{(delta or wye connections)}$$

Just as the voltage output of a generator at a certain frequency determines the size of the magnetic field and armature structure of steel, its current output determines the size of wire used in the machine. Therefore, the combination of magnetic structure (voltage) and wire (amperes) determines the size of the machine (kva).

Whenever the load voltage and current are in-phase, the load is absorbing real power, and the generator is delivering real power in kilowatts (kw). Real power is a measure of work or horsepower. In the event the phase current and phase voltage of a load connected to the generator are not in-phase, the load is said to be reactive. Reactive loads such as transformer banks, motors, and capacitors store power for a part of a cycle and deliver it back to the generator during the other part of the cycle (Fig. 2–10).

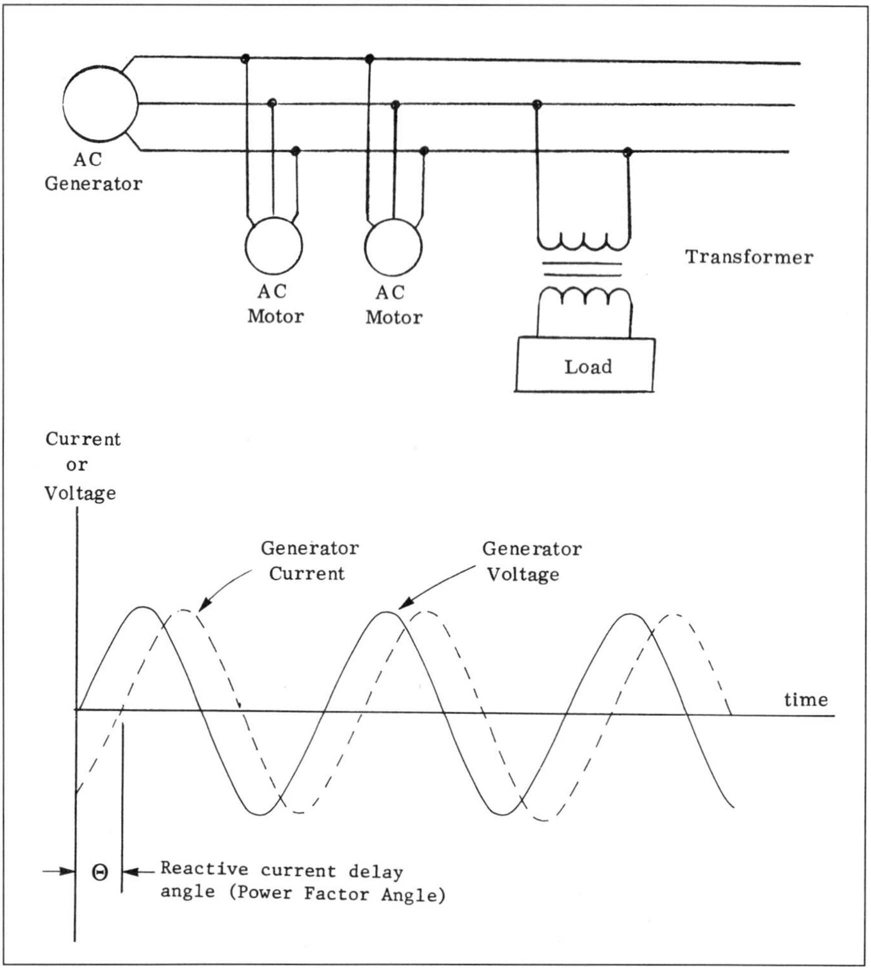

Fig. 2–10. Reactive loads and waveforms

Most electrical loads comprise both real and reactive components, and their vectoral sum is the load kva. The relationship of real and reactive power and total kva is:

kw = kva cos θ, kilowatts

where θ = angle between phase current and voltage
kw = real or useful power, kilowatts
cos θ = power factor

kvars = kva sin θ, kilovars
where θ = angle between phase current and voltage
kvars = reactive or useless power, kilovars

$$PF = \text{power factor} = \cos\theta = \frac{kw}{kva}$$

and $$kva = \sqrt{(kw)^2 + (kvar)^2}$$

Since cos θ or power factor can never be greater than unity, the maximum use of a generator is achieved when the kw equals the total kva. In most practical cases, the load kw is always less than the kva and, typically, gives a power factor rating of 0.7 for most machines in SCR systems.

A significant conclusion of diesel engines' and generators' ratings can be drawn from this analysis of a generating system:

A diesel engine produces only kw at its shaft (real power).

An AC generator delivers kva and its size is based on kva (apparent power).

Therefore, the diesel engine only delivers whatever kw is required by the generator and its load.

If a generator, for example, is delivering 1000 kva to a load at 0.1 PF, the diesel will be required to provide only 100 kw to the generator shaft. Similarly, if the PF were 1.0, the diesel would have to deliver 1000 kw.

Paralleling AC Generators

When two or more AC generators are paralleled to a common distribution bus, several criteria must be met to perform this function successfully. First, each generator must have the same phase sequence. If a generator has a timing phase sequence A-B-C, it must agree with other generator phase sequences for paralleling. If not, reversing any two of the three-phase leads will correct the phase sequence. Second, the voltage amplitudes must be approximately equal. Unequal voltages result in circulating currents between generators and not to the load. Finally, each corresponding phase to be paralleled must have coinciding phase angles. Any phase angle difference results in undesirable circulating currents between generators. Phase angle coincidence on a steady-state basis also assumes the same frequency for each generator.

If all generators are operating at the same frequency for each generator sequence, and their voltage amplitude and phase angles are identical, they can be paralleled without any circulating currents between them. If, however, they are paralleled 180° out of phase, excessive currents result and equipment or cable damage may occur.

After the generators are successfully paralleled to a common bus, they should share a proportionate amount of the load current. If the phases are essentially locked together at exactly the same phase, slight adjustment of the generator voltages will improve the balance of the load current from each generator as well as improving kva balance. Slight adjustment of the diesel engine speed adjustment will assist in improving the kw balance.

AC Voltage Control

Voltage control is achieved by sensing output voltage and current (sometimes) and adjusting the AC generator field power to provide the desired correction of AC voltage output. Synchronous generators are typically used in oilfield drilling applications, and the key to control is supplying DC excitation power to its rotating field. Generators of the type in Figure 2–11 have excitation systems that are either brush-type DC generators, static DC supplies, or brushless-type AC revolving armature exciters with rectified DC output to the field.

The exciter is the power control element in a voltage regulating system. Most drilling systems use a synchronous generator with rotating brushless exciter and static type voltage regulator for AC voltage control (Fig. 2–12).

The voltage regulator is the closed loop voltage controller for the system. By sensing the AC output of the generator and comparing it with a reference control signal, the current is regulated to the exciter field to maintain the desired output voltage. When the reference and feedback voltage from the generator output do not agree, an error signal initiates corrective action in the exciter to correct the output voltage from the generator.

An SCR-type voltage regulator provides a number of features: high-speed response to correct the AC voltage during generator load changes, regulation better than 1%, good voltage stability and low temperature drift, and rugged construction and high reliability.

Fig. 2–11 AC generator with brushless exciter (Courtesy Lima Electric Co.)

When a load is suddenly applied to an AC generator, its voltage drops due to the internal reactance of the machine. The recovery of this dip is a function of the voltage regulator response and the time constant of the generator/exciter. Removal of a load and the sudden rise and recovery of voltage is also a function of the same parameters.

Each manufacturer of voltage regulators provides several options of its

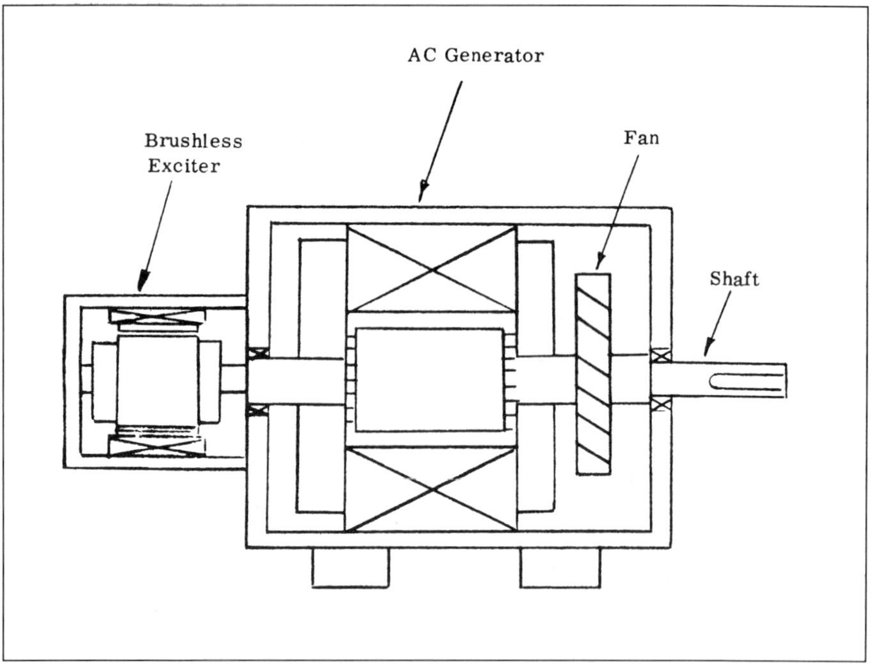

Fig. 2–12 Sectional view of synchronous generator with brushless exciter

standard line that define their performance:

1. Paralleling option, including current sensing transformer for load sharing, reactive power sharing, and automatic synchronizing.
2. Manual voltage control as required by United States Coast Guard.
3. Series boost option to provide overload excitation.
4. Protective features including detection and protective relaying of underfrequency, under/over voltage, ground faults, and overcurrent.

Paralleling generators to a common AC bus increases the kva capacity and improves the reliability of the electrical system. On most drilling rigs, two or more AC generators in parallel provide power for the various rig functions (Fig. 2–13).

The Institute of Electrical and Electronics Engineers (IEEE) has defined two methods of paralleling and controlling generator voltage. The first is *reactive droop compensation.* If two generators operating in parallel have circulating current flowing between them because one generator has an overexcited

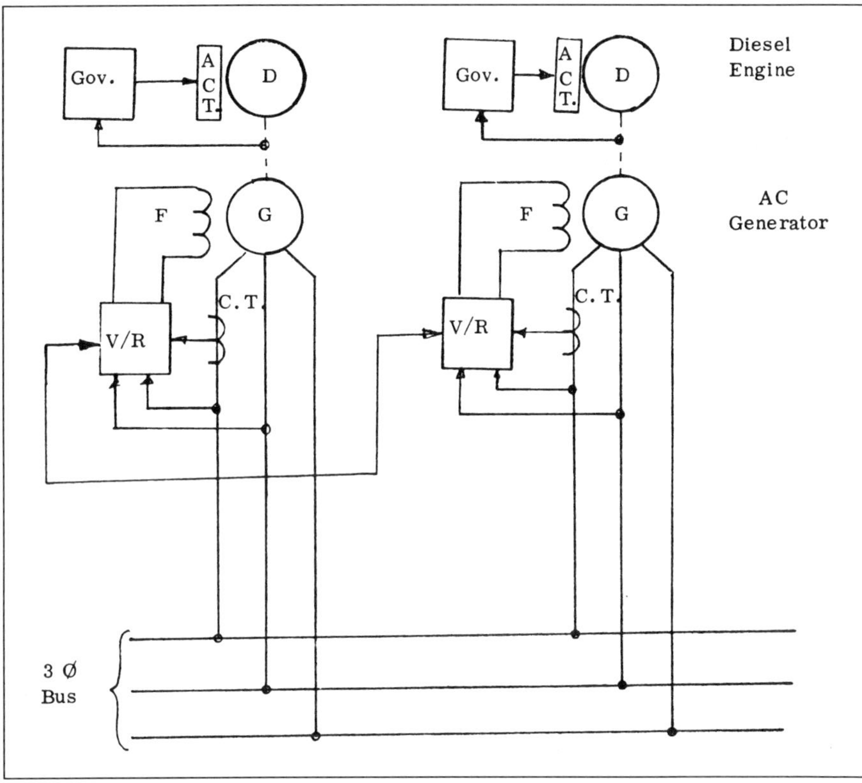

Fig. 2-13 Paralleled diesel-generator system

field, the current will appear lagging to the overexcited generator and leading to the other generator.

Using the reactive droop compensation technique, field excitation of one generator is reduced and the other field is increased to minimize the circulating current. An equal balance of reactive current circulating between the two units is left. Another effect of reactive current circulating is a reduction in the output AC voltage, the droop increasing with increased inductive current.

The second method is *reactive differential compensation,* more commonly referred to as *cross-current compensation.* This technique shares inductive loads without a drop in output voltage. Using circuitry similar to reactive droop compensation, current transformers from each generator output are connected in a closed series loop. When the loads are unbalanced, a voltage is developed in the series loop to create an adjustment in the generator excitation. Once the loads are balanced, the voltage in the series current transformer loop cancels to zero.

Generally, the voltage regulator and exciter are supplied with the generator manufacturer's equipment. In many cases, however, the voltage regulator must be defined by the generator manufacturer or an independent controls producer. To aid in this selection, the following specifications are required:

Exciter: voltage, kw, current rating, brush or brushless, static. Common voltages are 32, 63, 125 and 250 volts DC.

Generator: manufacturer, kva, voltage, PF, frequency, speed, phases, no load and full load excitation and field resistance.

Type and Quality of AC Power

Control Equipment

Most drilling rig systems have a complete control bay dedicated to each engine-generator set. Contained in this control cubicle will be diesel engine controls, consisting of the electronic engine governor (Fig. 2–14), voltage adjusting potentiometer, auto/manual selector switch, reverse power sensor and relay, underfrequency/overvoltage sensor and control, generator circuit breaker, instruments and meters, and indicating lamps.

Manufacturers will incorporate some of these functions in a completely self-contained module resulting in a voltage regulator/governor control. This is sometimes referred to as the "AC Module" by the manufacturers.

The governor closely controls the engine speed by comparing the actual speed to that set on the speed-adjusting potentiometer. As the load changes, the governor produces a voltage to the fuel throttle, or actuator, to increase or decrease the diesel fuel flow and maintain the desired speed. The speed sensor, located at the engine, is usually a magnetic pickup producing pulses of voltage to the governor.

Additional inputs to the governor are the voltage and current sensing transformers from the generator output for load sensing and a DC voltage supply (usually 24 volts DC) for the governor electronics. The governor switch has three positions: *shutdown*, to move the actuator to minimum fuel flow position; *idle,* to maintain no-load idling speed; and, *run,* to operate at full-rated speed.

The generator controls include the voltage regulator for voltage sensing and control, output power for the exciter field (if rotary), and parallel operation.

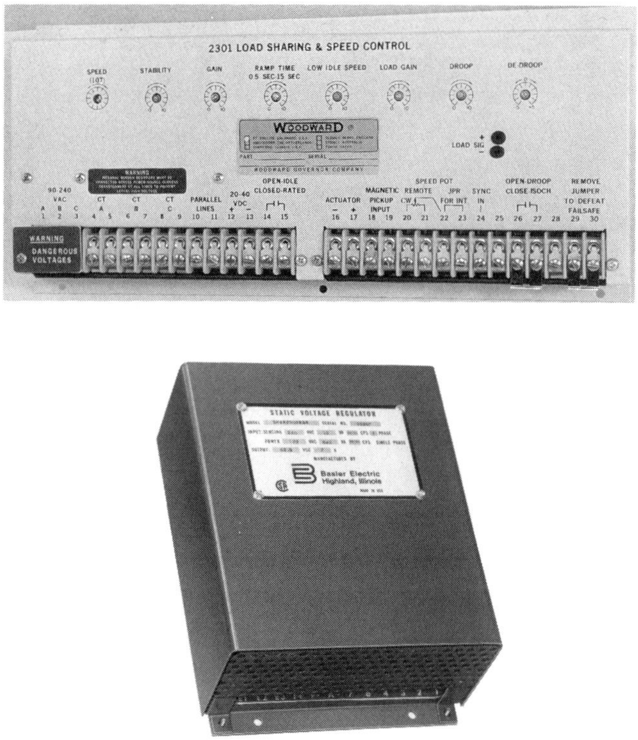

Fig. 2–14 Electronic governor and voltage regulator. (Courtesy of Woodward Governor Company and Basler Electric Co.)

Sensing the output generator voltage and comparing it to the desired reference setting on the voltage-adjusting potentiometer changes the exciter and generator field currents as required to maintain constant AC voltage.

Parallel Operation

When two or more generators are in parallel, the voltage regulators sense voltage and current to maintain equal voltages and minimize circulating current between generators. Presetting the voltage-adjusting potentiometer to a minimum circulating current minimizes the correction required by the regulator. When a generator is removed from the line and is no longer in parallel with the other generators, it will short-out its current transformers to prevent false signal transmission of the crosscurrent compensation network.

Should one of the generators receive more power than it delivers, it is, in effect, receiving reverse power, which may overspeed the diesel. Once this occurs, the reverse power sensor and relay open the generator circuit breaker and remove the generator from the AC line.

If desired, the voltage of each generator can be controlled manually or automatically by proper selection of the manual/auto select switch. A switch can also manually or automatically synchronize the generator to the AC bus.

Should an underfrequency or overvoltage condition occur beyond preset levels, the generator regulator will be turned off and the generator removed from the line. Also, if a fault such as an AC ground occurs, an alarm will be activated or the generator will be removed from the AC bus.

The generator AC circuit breaker operates continuously under the full-load kva of the generator and functions to protect the generator. Engine protection is a separate function determined by its kw rating; the circuit breaker is not sized to protect the engine, but to protect the generator. Fault current analysis determines the interrupting capacity rating of the circuit breaker. Most circuit breakers have adjustable current trip settings in addition to an undervoltage release sensor and auxiliary contacts for additional control.

Meters used to monitor the generator typically measure voltage, current, frequency, power factor, kilowatts, kilovars, kilovolt-amperes, and engine temperature. Lamps indicate generator and circuit breaker status, sync indication, and ground (Fig. 2–15).

Voltage Quality

The ideal AC system has a pure sine wave, zero generator impedance, stable amplitude under all load conditions, and excellent frequency stability. Of course, the cost of such a system is prohibitive and impractical, but it does provide a reference from which a practical AC generating system can be measured. On drilling rigs, the generating system should provide all anticipated loads reliably, with enough margin to allow some additional loads in the future.

Several quality generating and control systems have steady-state voltage and frequency regulation of 1% or less and a waveform with minimum distortion under linear load operation. Adding SCR power equipment on the AC bus increases the electrical distortion of the waveform with "commutating notches" and switching transients. These notches usually will not affect normal AC

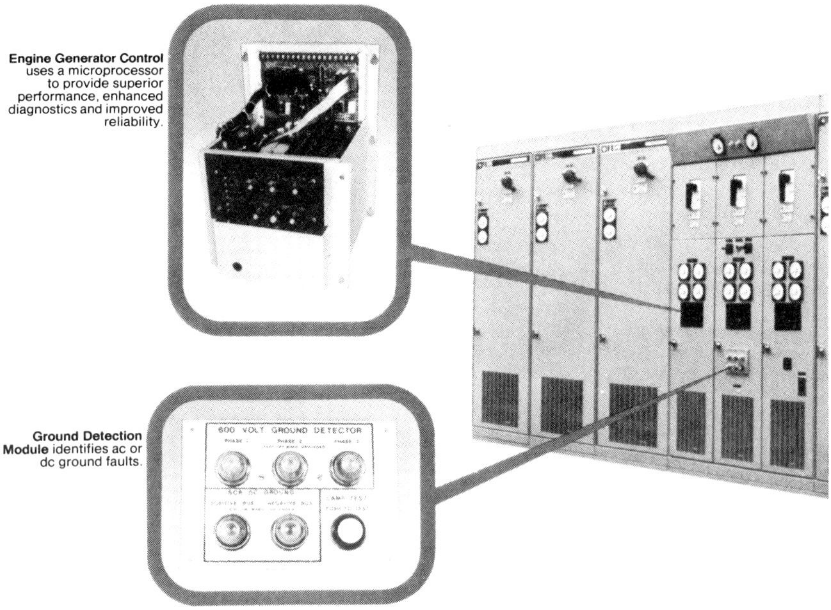

Fig. 2–15 Generator control cubicle for a land drilling rig (Courtesy of Ross Hill Corp.)

motors operating from the bus but may influence control equipment using the AC without any additional filtering or buffering.

As long as the AC bus voltage and frequency remain within the operating limits of the equipment connected to the AC line, the system should be considered acceptable in most applications.

Emergency Power Generation

Some equipment on the rig may not tolerate a power outage without equipment damage or loss of critical information. Such equipment includes onboard computers or instrumentation on offshore vessels. Other equipment can tolerate a power outage, but an alternate source of AC power may be required for certain equipment before the main bus power is restored.

Uninterruptible Power Supplies

Uninterruptible power systems (UPS) supply continuous AC power to a load even during a primary power outage. This equipment can be rotary or static, depending on the performance desired. A rotary system uses an AC motor, DC motor and AC generator coupled together on a common shaft. Normal power flow is from the AC bus to the AC motor driving the AC generator

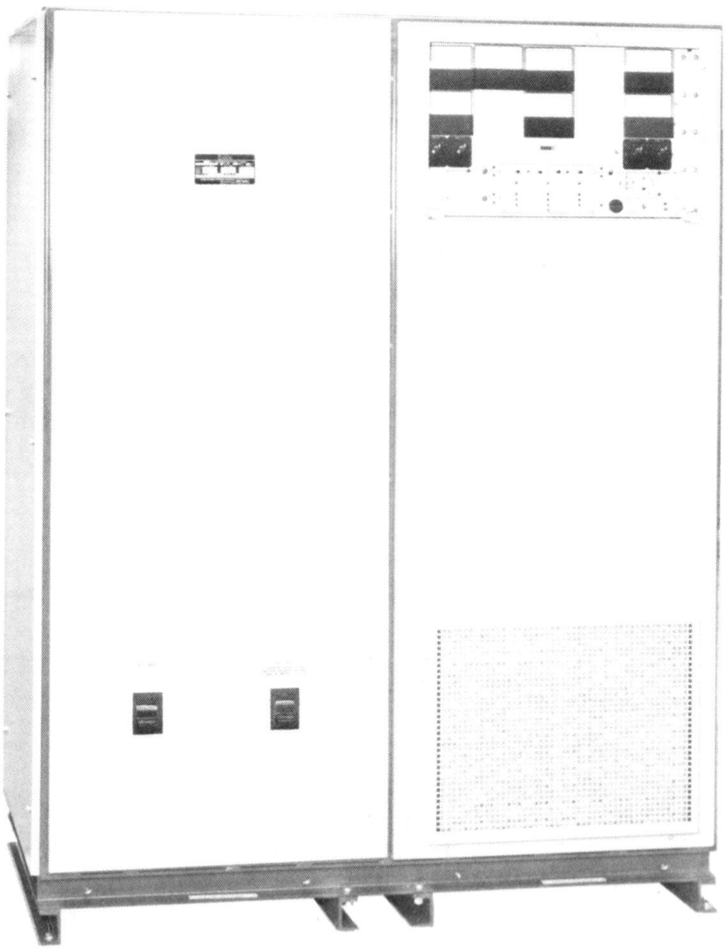

Fig. 2–16 Static UPS system, 30 kw rating (Courtesy Exide Conversion Division, Inc.)

that supplies AC power to the critical load. The DC motor runs at no load with no power applied to it. In the event of primary AC power outage, the battery switch and regulator energize the DC motor to continue driving the AC generator for continuous power. When primary power is restored, the system operates again from the AC supply and motor (Fig. 2–16).

A completely static UPS has no moving parts, is highly efficient, and provides continuous power at the time of primary AC power loss. The battery automatically assumes the DC power required by the inverter until primary AC is resumed. Several quality manufacturers have produced this product for many years at power levels up to 250 kw and higher.

Diesel Engine-Generator Sets

If considerable power is needed and the load can tolerate a short power outage, a standby diesel-generator set can be used on the rig. Start-up of the emergency generating set occurs as soon as it is determined that the main AC bus is down. After the set is operating at full speed, certain noncritical loads are isolated from the AC bus, leaving only those items deemed critical and within the capacity of the AC generator. When this load isolation occurs, the emergency generator is placed on the line to supply the AC power.

Once main AC power is produced again, the emergency set is taken off the line and all normal functions are restored. A typical system of this type is shown in Figure 2–17 along with a typical 400-kw system in Figures 2–18 and 2–19.

Fig. 2–17 Emergency diesel-generator skid assembly

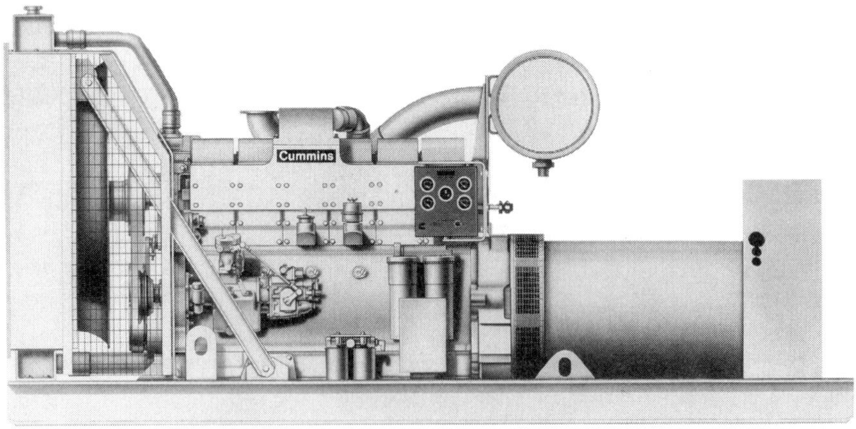

Fig. 2–18 Prime power or emergency 450 kw diesel-generator set (Courtesy Cummins Engine Co. Inc.)

Fault Current Analysis and System Protection

Protecting the complete AC generating and distribution system, including the engine, generator, circuit breaker, and the interconnecting cable and bus bars, prevents costly damage resulting from faults that can occur. Coordinating the generator's fault currents during circuit breaker trip ratings is important. The ability to clear the fault quickly is one of the main requirements of the circuit breaker and knowledge of the maximum fault current is needed to select the circuit breakers in the system.

Performing a fault analysis of the generating system requires knowledge of: 1) the generator voltage and internal impedance; 2) series transformer impedances; and, 3) active motor loads. Prior to the fault calculation, information on the number of AC generators and their ratings, their sub-transient reactance (X"d) and the transformer impedances must be known.

An example of a fault calculation is given for the system in Figure 2–19. The fault is assumed to occur at the load side of circuit breaker CB7 (Fault #1) with the circuit values as shown. The fault current to determine interrupting currents through CB7 is as follows:

Available generator equivalent impedance:

$$\%Z_g = \frac{13\%}{3} = 4.33\%$$

Transformer equivalent series impedance:

$$\%Z_{T1} = 6\%$$

Total equivalent impedance:

$$\%Z_T = 4.33\% + 6\% = 10.33\%$$

Fault kva:

$$\text{Fault kva} = \frac{\text{base kva}}{Z_T} = \frac{1.875}{0.1033} = 17.480 \text{ kva}$$

Fault current at 600 volts:

$$I = \frac{(\text{kva})(1000)}{(V)(1.73)} = \frac{(17.480)(1000)}{(600)(1.73)}$$

$$= 16,840 \text{ amperes symmetrical}$$

Therefore, the interrupting capacity of CB7 must be greater than 16,840 amperes to assure successful interruption of this fault current.

Following a similar approach, Fault #2 occurring on the 600-volt AC bus will produce a total bus fault current of 43,300 amperes and will divide equally through circuit breakers CB1, CB2, and CB3 at a current level of 14,430 amperes each.

Lightning Protection

Lightning Character

Any equipment exposed to the environment and extending vertically into the air is a candidate for lightning strikes; e.g., a land drilling rig whose mast may extend several hundred feet from the ground surface. The taller a structure stands, the greater the probability of it being struck by lightning.

The energy in a typical lightning flash can be measured in several ways. Since it is basically a movement of electrons from one location to another (cloud to rig), it is best described in terms of current flow.

Lightning's energy is the current squared multiplied by the interval of time during the flow. Therefore, a typical flash would be:

$I^2\,(T_f - T_i) = 55,000\ \text{amp}^2 - \text{seconds}.$

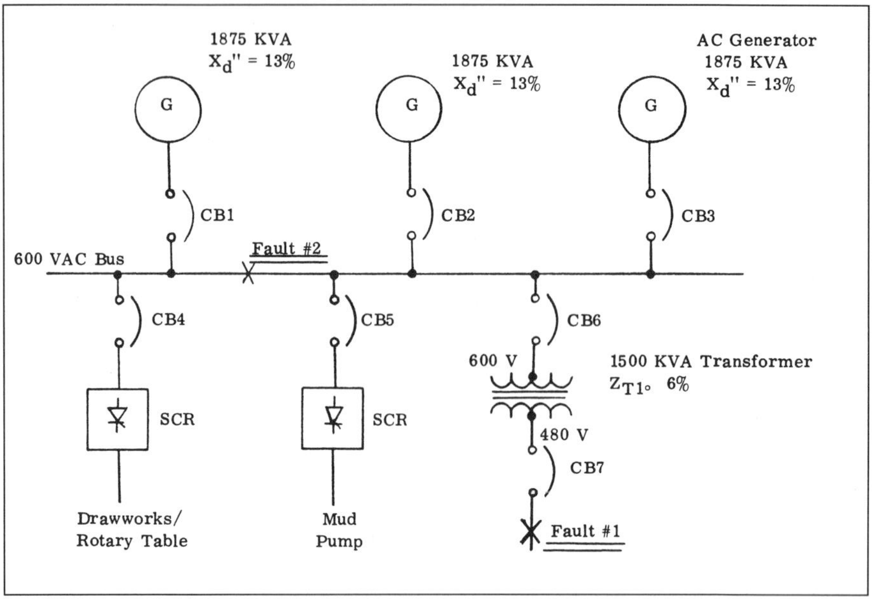

Fig. 2–19 Electrical one-line diagram for fault current analysis.

A typical flash consists of 10 strokes of lightning and lasts for approximately 0.3 seconds. Each stroke averages 30,000 amperes peak and is a current pulse whose rise time is 2–6 microseconds and whose decay time is 50–200 microseconds. The destructive energy of lightning is its energy intensity— high currents in a very short time. At times, lightning may occur as a steady current of 100–200 amperes from cloud to earth and can be responsible for causing forest fires or igniting combustible gases.

Effects of Lightning

Lightning surges, or currents, can hurt or kill personnel and damage rig equipment in two ways.

Since the lightning strike is essentially a constant current effect, it flows from the charged cloud to the rig mast down to the earth. Its conducting path exhibits an equivalent resistance to this flow and develops a voltage drop along the way to ground.

For example, a ground system with 4 ohms of resistance hit by lighting surge of 30,000 amps develops a voltage of V = IR = 30,000 x 4 = 120,000 volts. As this current flows into the ground radially from the rig, it develops smaller voltage drops at the ground surface which could amount to 5,000–10,000 volts between a man's feet standing on the ground. A similar voltage developed along the mast could damage motors, transformers and other electrical equipment.The other electrical effect exhibited by lightning is its ability to induce current flow in electrical conductors because of the high magnetic field surrounding the main current flow. With the mast appearing like a transformer primary, other rig components become secondaries with current loops to produce high voltages, shock, and damage.

Lightning Protection

Using this knowledge as a precaution, the rig designer and operator can install protective devices for electrical equipment and design features to minimize lightning damage. The conducting path from the top of the mast to ground should be as low in resistance as possible, all metal structures should be bonded together tightly, and the mast should be well-secured to ground. Electrical gear directly exposed to lightning (such as transformer primaries) can be protected with lightning surge arrestors, spark gaps, or metallic oxide varistors (MOVs). The extent of this protection depends on the area being drilled and its incidence of thunderstorms. A typical rig in a moderate thunderstorm area has a 6% chance of being struck by lightning in a given year and a 25% chance of being struck over a five-year period.

Bibliography

Anderson, J. G. "Lightning to Drilling Rigs." Paper presented to the 1979 Drilling Technology Conference of IADC, March 1979.

Hudson, Ralph G. *The Engineers Manual*, 2nd Edition, New York, John Wiley & Sons, 1957.

Jaeschke, Ralph L. *Controlling Power Transmissions*. Cleveland: Renton/IDC Inc., 1978.

Newell, A. B. *Diesel Engineering Handbook*. 9th Edition, New York: Diesel Publications Inc., 1955.

3 ELECTRIC POWER DISTRIBUTION AND CONTROL

After the AC power is generated and controlled, its distribution throughout the rig is important for successful operation. Selecting the proper circuit breaker and cabling and understanding the character of the individual loads are imperative. Guidelines have been established by IADC, API, NEC, NEMA, IEEE, ANSI, USCG, and ABS for proper selection criteria.

These guidelines should be selected for specific applications since land rigs and offshore rigs differ in the requirements imposed on them. Classification of hazardous and nonhazardous areas on drilling rigs has been defined by the API's Recommended Practice RP500B and should be referred to in all cases.

The System AC Bus

The reliability and rating of the electrical system are predicated on the integrity of the AC bus. This bus is established by the paralleled AC generators, and its rating at any given time depends on the power available from the diesel-generator (D-G) sets. If four D-G sets are on the rig, but only two are operating to the AC bus, the bus rating at that moment is the sum of the two operating D-G sets.

Anticipating load requirements improves the operating efficiency of the rig through fuel savings and maintenance costs. When conditions exist that do not require peak power loads, a minimum number of D-G sets can be connected to the AC bus. Steady drilling with light mud pump and hotel loads is one condition where peak power requirements are low. On the other hand, operating mud pumps and hotel loads while tripping requires high AC power because of pipe string load and traveling block acceleration requirements. Knowing these power conditions allows the rig superintendent or tool pusher to use only the required D-G sets.

Fig. 3–1 Three-generator control system (Courtesy Baylor Co.)

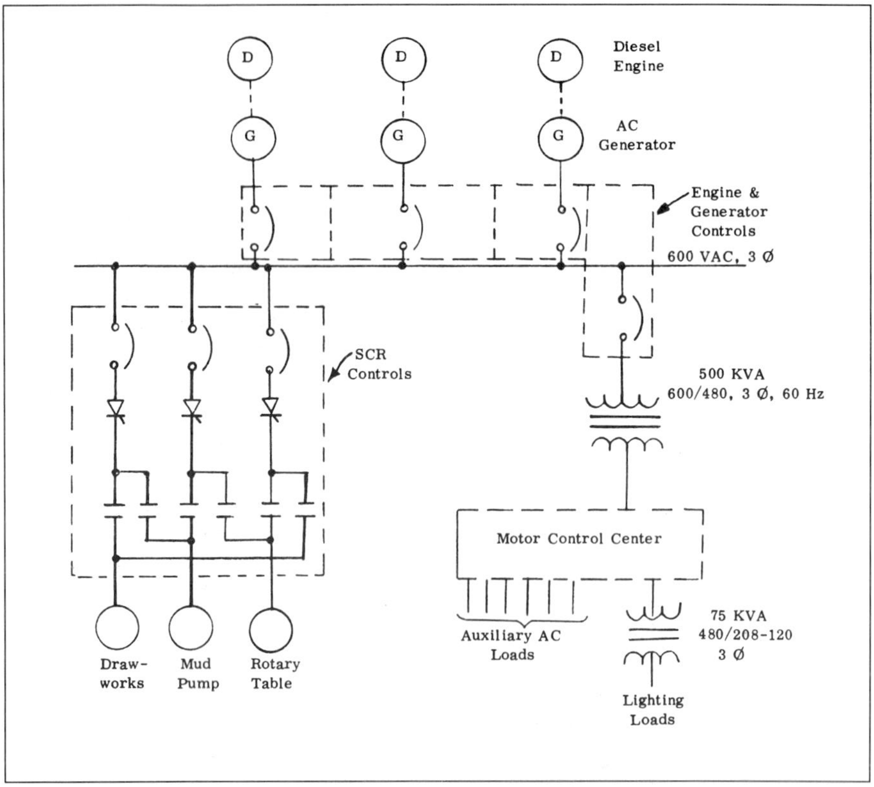

Fig. 3–2 Typical electrical one-line diagram of a land rig system.

Operating personnel should know the effects that accelerating heavy loads, such as drawworks tripping, has on the AC bus. If inadequate D-G sets are used, the excessive current requirements may stall the engines and trip them off the line due to underfrequency, or the transient voltage drop on the generators may operate the undervoltage detector.

In either case, the required drilling function is not achieved and a system restart is necessary. Adequate D-G sets must be kept on the AC bus to meet peak power and kva demands of the loads. A rig set of three generator controls is shown in Figure 3–1, a typical one-line land rig power distribution system is shown in Figure 3–2, and an offshore one-line is shown in Figure 3–3.

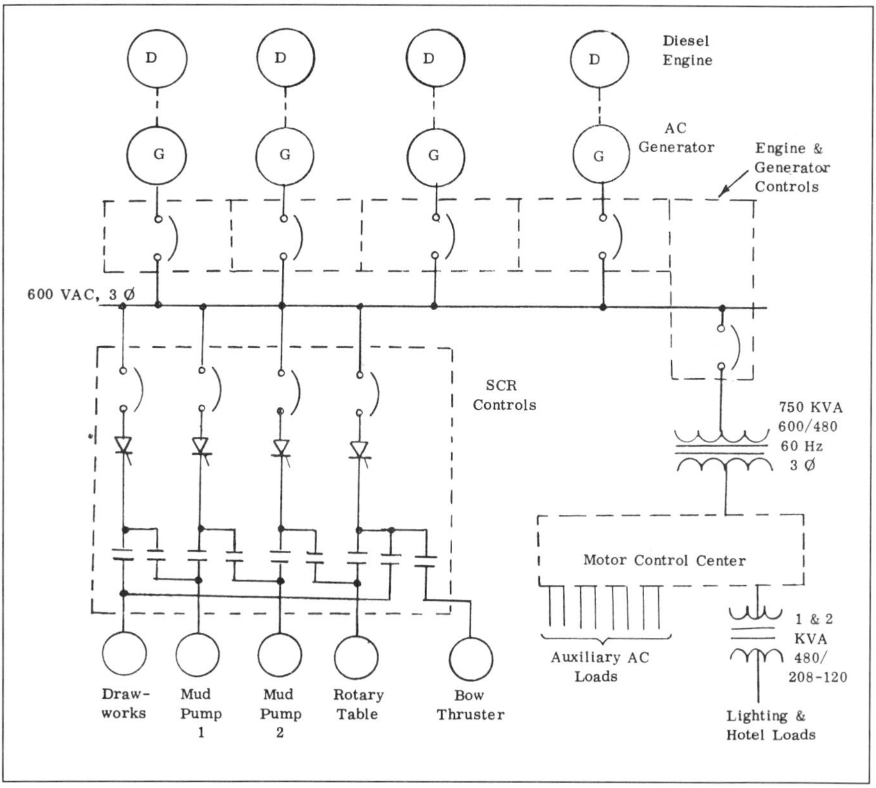

Fig. 3–3 Typical electrical one-line diagram of an offshore drilling system.

AC Switchgear

Most drilling rigs use power at levels of 600, 480, 208, or 120 volts and as such are classified "low voltage" by the major electrical standards groups. The classification allows live circuit components to have a lower spacing between them and interrupting ratings of circuit breakers and switches lower than medium- or high-voltage circuit components.

Low-voltage AC switchgear distributes power from the main AC bus and interrupts branch circuits without shutting down the generators or disrupting the main AC bus. This switchgear usually consists of circuit breakers to control the power to large transformers, AC motors, rectifiers and SCR supplies, and other heavy-duty loads or transfer switches from main to emergency power buses.

On land drilling rigs, a main circuit breaker powers the 600–480 volt, three-phase transformer used for auxiliary AC motor loads. This transformer may be as small as 250 kva or as large as 1000 kva and the circuit breaker must be rated for both its full load and inrush characteristics. Other circuit breakers must be provided for the 600–208/120 volt lighting transformer which is rated between 75 kva and 150 kva. AC circuit breakers for the SCR drives should also be provided.

Offshore drilling rigs typically require greater hotel and lighting power and greater auxiliary AC motor loads. Circuit breakers for these units must have higher ratings and in some cases, because of USCG and ABS requirements, have a drawout frame assembly for disconnecting from a "hot" AC bus.

One of the main recognized standards for switchgear is the ANSI/IEEE standard for "Switchgear Assemblies Including Metal-Enclosed Bus," identified as ANSI C37.20 and IEEE Standard 27-1974. These standards recognize switchgear arrangements as metal-enclosed power switchgear, metal-enclosed bus, and switchboards.

The metal-enclosed switchgear and buses provide protection by enclosing all primary and secondary switching components, transformers, instruments,

Fig. 3–4 4160-volt switchboard for offshore rigs (Courtesy Baylor Company)

fuses, relays, and their wiring in a cubicle. Access to these components is gained through doors or panels. Switchboard assemblies consist of panels with externally mounted instruments, switches, and other components accessible to the operator-user (Fig. 3–4).

On drilling rigs with DC-drive motors, DC switchgear energizes the motors from an SCR or DC generator source and alters assignments of the motors from one source to another. Dynamic braking and reversing are also accomplished with DC contactors and their associated relay logic (Fig. 3–5).

Fig. 3–5 DC switchgear assembly (Courtesy Baylor Co.)

In some cases, AC power is generated higher than 600 volts, usually 4160 volts. In such applications, the switchgear is metal-clad, which is characterized by live parts enclosed in grounded metal compartments, removable primary switching devices, insulated bus bars, mechanical interlocks for safe operation, and isolated instruments from the primary circuit.

Large AC motors designed for this voltage level reduce conductor size and current levels required.

Transformers

Converting AC power from one voltage level to another and providing electrical isolation is the primary purpose of transformers on drilling rigs. If the primary power developed by the AC generators to the system bus is 600 volt, three-phase, 60 Hz, transformers will be required for the following functions:

LOAD	VOLTAGE REQUIRED
Lighting and hotel loads	120/208 volt, three phase, four wire
Motor control center	480 volt, three phase
AC thruster motors	Varies with manufacturer
Control functions	120 volt, single phase
Instrumentation	120/24 volt, single phase

Power transformers on land rigs are usually provided up to 750 kva for most applications (primarily for auxiliary motor loads) and can be up to several thousand kva for offshore systems (primarily for thruster motors). Except in special cases, all power transformers above 25 kva are typically three-phase, dry-type transformers, either self-cooled or fan-cooled in a steel enclosure. Where conditions permit, liquid-filled, self-cooled transformers are used at high kva levels.

Land rig applications use high-quality transformers manufactured with ANSI and NEC standards. Offshore systems, in addition, use dry transformers which are properly vacuum-impregnated and further protected against the environment. Copper is normally used for the wire coils (vs aluminum) with Class H temperature-rated insulating materials. Nameplates on transformers should be permanently mounted with proper rating and manufacture identification in accordance with ANSI C57-12.70 standards.

Mounting and installing transformers is usually done in accordance with applicable local electrial codes and per Article 450 of the *National Electrical Code*.

Care should be exercised when mounting transformers in hazardous areas where the transformer is exposed to chemicals, corrosive materials, or explosive gases. These environments may require stainless steel enclosures or explosion-proof enclosures in accordance with NEMA 7, Class 1, Division 1 requirements.

Transformers are usually connected to terminal blocks for low kva-rated transformers and bolted studs or bus bar for large kva units. Proper connections at high current points must be made to prevent overheating and failure. Proper spacing should also be maintained for the voltages involved between adjacent connections.

Transformers operating from three-phase systems can have different winding configurations based on a wye or delta connection. Most are based on whether the feeder system is three-wire (delta) or four-wire (wye) and on whether the loads are designed for three- or four-wire terminations. Five configurations of primary to secondary windings are the wye-wye, delta-delta, delta-wye, wye-delta, and open-delta secondary.

Limitations of a delta winding are the inability to obtain line-neutral voltages without a separate grounding transformer and the close voltage balances of each winding to prevent circulating currents in the closed delta. Wye-wye windings can produce third harmonic currents which can be eliminated with a wye-delta or delta-wye transformer. The open delta transformer can be used in the event one winding of a closed delta transformer fails to open, but the transformer rating is reduced to 57.7% of the original kva rating.

Power transformers on rigs typically have the ratings and characteristics of Table 3–1.

Table 3–1

Typical Drilling Rig Transformer Ratings

Primary Voltage	Secondary Voltage	kva	Coil Material	Weight (lbs)
600	480	250	Copper	1500
600	480	500	Copper	2800
600	480	750	Copper	4100
600	480	1000	Copper	4700
600	480	1500	Copper	8000
600	480	2000	Copper	8500
600	208/120	75	Copper	640
600	208/120	112	Copper	780
600	208/120	150	Copper	1000

Notes:
1) Ratings are at 60 Hz, three phase
2) Ambient temperature range: –30° C to 40° C
3) 600-volt class, outdoor, dry-type, 220° C insulation class

Motor Control Center

Distributing power from the 600/480-volt transformer to auxiliary AC motors on the rig is usually accomplished by a combination of circuit breakers, contactors, and overload relays connected as full-voltage starters that constitute a motor control center. Each starter is defined by NEMA in various classes, with each class covering a range of horsepower at a certain operating voltage. Most auxiliary rig motors are rated at 480 volts, three phase, 60 Hz; they range in size from NEMA size one through five. A breakdown of starter and motor ratings at 115, 230, 460 and 575 volts is shown in Table 3–2.

TABLE 3–2
Full=Voltage Motor Starters—NEMA Ratings

NEMA size	Horsepower at three-phase voltages 230 V	460 V	575 V	Current Ratings (amperes)
00	1.5	2	2	9
0	3.0	5	5	18
1	7.5	10	10	27
2	15.0	25	25	45
3	30.0	50	50	90
4	50.0	100	100	135
5	100.0	200	200	270
6	200.0	400	400	540

Three-Phase AC Induction Motor Full Load Currents
(Nema B Design motors at 1725 rpm)

Nominal Motor Voltages

HP	115 V	230 V	460 V	575 V
0.5	4.0	2.0	1.0	0.8
1	7.2	3.6	1.8	1.4
3	-	9.6	4.8	3.9
5	-	15.2	7.6	6.1
10	-	28.0	14.0	11.0
15	-	42.0	21.0	17.0
25	-	68.0	34.0	27.0
50	-	130.0	65.0	52.0
75	-	192.0	96.0	77.0
100	-	248.0	124.0	99.0
150	-	360.0	180.0	144.0
200	-	480.0	240.0	192.0

Each manufacturer of motor control centers (MCCs) has standard section heights and widths to provide a uniform lineup. The combination of sections makes up an MCC and can comprise from one to 20 sections or more. Within each section are starters of various NEMA sizes, depending on the auxiliary AC motor ratings. To utilize each section properly, starters are assigned "space factors" and each NEMA size has its particular space allocation (Fig. 3–6).

Fig. 3–6 Typical motor control center of three sections (Courtesy Square D Company)

A drilling rig has many motor control functions, and each must be fully defined before design of the MCC. The best way to create the MCC configuration is to list each motor, heater, or other three-phase load with its rating and characteristic. From this list an MCC definition such as in Table 3–3 can be made. Both API Standard 542 and NEMA detail MCC specification standards.

Table 3–3
Typical Motor Control Center List for Land Rig—480 Volts

Function	Full voltage nonreversing HP starter size	Control	Circuit breaker trip/frame
Spare	100/4	PBSS	
Mud mix #1	75/4	PBSS	
Mud mix #2	75/4	PBSS	
Desanding	75/4	PBSS	
Desilting	75/4	PBSS	
Air compressor #1	40/3	HOA	
Air compressor #2	40/3	HOA	
Super charger #1	40/3	HOA	
Super charger #2	40/3	HOA	
Spare	40/3	HOA	
Water pump #1	25/2	PBSS	
Water pump #2	25/2	PBSS	
Spare (2)	25/2	PBSS	
Water well pump	10/1	PBSS	
Degasser	10/1	PBSS	
Mud agitator #1	10/1	PBSS	
Mud agitator #2	10/1	PBSS	
Mud agitator #3	10/1	PBSS	
Mud agitator #4	10/1	PBSS	
Mud agitator #5	10/1	PBSS	
Mud agitator #6	10/1	PBSS	
Mud pump blower #1	10/1	HOA	
Mud pump blower #2	10/1	HOA	
Drawworks A blower	10/1	HOA	
Drawworks B blower	10/1	HOA	
Drawworks oiler	10/1	HOA	
Spare	10/1	HOA	
Elmagco brake			100/100
Wire line unit			70/100
Power house A/C			70/100
BOP			50/100
Allenvator			40/100
Shale shaker			15/100
Air heater			15/100
Spare			15/100

Notes: Enclosure is NEMA 1 with type B wiring.
Bus is copper and braced for 22,000 amps RMS.
Container horizontal ground bus
PBSS = Pushbutton, stop/start
HOA = Hand-Off-Automatic

Lighting and Hotel Loads

Fluorescent and incandescent lighting can be used for offices and rooms, mast illumination and signaling, areas requiring local lighting, and large areas. Safety and job performance are the primary reasons lighting is used on rigs. Hotel loads include galley appliances, workshop loads, entertainment, and communication devices.

Power for these loads is derived from the 120/208-volt transformer through the lighting distribution panel designed in accordance with NEMA standards. A typical panel consisting of 42 single-pole spaces rated for 225 amps is shown in Table 3–4.

Table 3–4
Typical Land Rig Lightning Panel Loads and Ratings

Functions	Amps/phase
Spare	30/1
Spare	30/1
Spare	30/1
Spare	30/1
Spare	30/1
Spare	30/1
Generator #2 heater	30/1
SCR indicator lamps control power	30/1
SCR house receptacles	30/1
SCR house lights	30/1
Engine skid #2 lights	30/1
Engine skid #1 lights	30/1
Tank #2 lights	30/1
Tank #2 lights	30/1
Tank #1 lights	30/1
Generator #1 heater	30/1
Mud pump #2 heater	30/1
Mud pump #1 heater	30/1
Drawworks heater	30/1

Marine Sanitation Devices

Sewage treatment on offshore rigs is usually accomplished by marine sanitation devices (MSDs). The size of these units is based on the gallons per day of liquid to be processed and the maximum number of personnel being handled on the rig. Each of these MSDs requires one or more electric pumps to maintain proper flow-through and processing rates.

Most MSD motors are low horsepower, single-or three-phase sizes. Power distribution for these units is normally through the lighting panel.

Drives and Controls for Drilling

Typical drilling functions, such as drawworks, mud pumps, and rotary tables, receive electrical three-phase power directly from the generator bus, and control it through an SCR converter for conversion to mechanical power through the DC motor. Distributing power from the system AC bus to the SCR units is typically achieved with a short extension of the bus bars directly to the SCR units from the generator control unit. In some cases, a run of appropriately rated insulated cable from the AC bus to the SCR units is required.

Just as each generator is protected with a separate circuit breaker, each SCR unit uses an input circuit breaker for fault protection and isolation from the three-phase AC. These circuit breakers trip in the event of an SCR/drilling function failure or overload without interfering with the system AC bus.

The output of the SCR unit is controlled DC voltage with a current magnitude proportional to the torque on the drilling motors. Usually, the distance separating the SCR control and DC motors is 50–300 ft, and proper cable type and size must be considered. On a land drilling rig, the cables are terminated to the end of the power house and run to the individual DC motors, while direct connections from the SCR bay to the motor are made on offshore rigs.

Control cables are run from the SCR bays and generator control bays to the driller's console located on the drill rig floor. Since this multiple-conductor cable contains the control and alarm signals to operate the drilling functions, it should be adequately terminated and protected against hazards on the rig. Laying unprotected cable on the rig floor should be avoided. Preferably, the cable should be run in conduit, be armor-protected, or placed in cable trays on the shortest route possible around the rig.

The size and type of all cables should follow the appropriate sizing requirements outlined by NEC and IADC and the controls manufacturers. The varying power factor required by the SCR unit should be considered in selecting the proper size cable from the AC generators to the SCR units as it will vary depending on the drilling motor speed. At low speeds, the SCR input power factor may be as low as 0.2–0.4, but with moderate to high speeds, the power factor will be 0.7–0.9.

Propulsion, Jacking and Other Drives

Semisubmersibles, jackups, and drillships use thrusters for main propulsion, propulsion assist, docking maneuvers or dynamic positioning. These thrusters can be powered by SCR variable-speed drives or operated with constant-speed AC motors to variable-pitch propellers.

Selecting the power cable from the power source to the motor should be made after considering maximum starting load and overload conditions. The drive motors should be protected to avoid downtime, equipment damage, and potential fire hazards.

The thruster console located on the bridge of the vessel controls the motor speed or propeller pitch. Control cables from the console to the power control unit should be carefully routed and protected to avoid damage and loss of control.

Jacking drives on jackup rig systems usually consist of electro-hydraulic units, constant-speed AC motors, or variable-speed SCR DC drives. Control of each drive is usually located at one control area to ensure coordinated control and level jacking. At least 16 drives provide the jacking power (Fig. 3–7).

Radio Frequency Interference Protection

Equipment such as SCR power converters, DC motors, and DC generators can produce high-frequency electrical noise or signals that are in the radio frequency bands. If communications equipment is used on the drill rig, these noise signals may interfere with transmission and reception of radio signals unless precautions are taken in cable routing and proper grounding techniques. Therefore, all signal-sensitive electronic cables should be run in separate conduits or trays from noisy power wiring to minimize coupling. Cables carrying electronic data or signals should be shielded, twisted, pairs with the shield properly grounded at one end only. In some instances, electromagnetic shielding of radiating sources may be required in addition to filtering.

Fig. 3–7 Electric motor driven jacks for jackup (Courtesy National-Oilwell)

Wiring and Cabling

Selecting the proper wire and cable for a particular application is important in reducing downtime and maintenance and for ensuring proper equipment operation. One should determine the maximum current and insulating voltage required by the particular circuit and the maximum voltage drop permissible by the load equipment. Also, determine the ambient temperature and environment to which the cable will be subjected, whether single- or multiple-conductor cable is required, and how many cables are to be run together in a conduit or cable tray. Finally, determine if the system is a land rig or offshore system; USCG and ABS rules will apply on most offshore vessels and applicable wiring codes should be reviewed. With this information, select the insulation type and wire size that meets the appropriate codes and standards at minimum cost.

Bibliography

ANSI/IEEE Standard for Switchgear Assemblies Including Metal Enclosed Bus. New York: Institute of Electrical and Electronic Engineers, Inc., 1974.

API Standard Practice for Design and Installation of Electrical Systems for Offshore Production Platforms. 1st Ed., Washington, D.C.: American Petroleum Institute, July 1978.

National Electrical Code. Boston: National Fire Protection Association, 1990.

4 ELECTRIC POWER UTILIZATION

Using electrical power effectively once it is generated and distributed is the primary purpose of the rig electric system. More electrical functions are being added to rigs and the effect these loads have on the system should be understood. Parameters such as starting or inrush current, power factors, load efficiency, accelerating currents for motors, duty cycle, and the operating environment contribute to the system bus performance.

Ward-Leonard DC-DC Drives

Ward-Leonard DC-DC drives on drill rigs usually consist of a diesel engine coupled to a DC generator operating at essentially constant speed. The DC output of the generator is controlled by varying its shunt field excitation, and its variable generated power is delivered to the DC drilling motor.

This system is essentially dedicated for a single purpose. Any load changes caused by drilling activity reflect directly on the DC motor-DC generator and engine and rarely interfere with other rig functions. The engine and DC generator must have adequate capacity to supply full load and accelerating current under all load conditions over the operating speed range.

SCR AC to DC Drive Alternatives

Ward-Leonard DC-DC drives have been replaced on many drill rigs with a silicon-controlled rectifier (SCR) solid- state speed-control system. One of the main advantages of this system is its ability to operate from the common AC system bus that supplies other rig functions. Using SCR systems to convert the fixed-voltage AC into variable voltage DC for the DC drilling motors eliminates the need for dedicated diesel-generator sets for each drilling function. Several SCR units can operate from the system AC bus if the total kw power and kva available from the engines and AC generators exceeds the demand required. A general classification of these systems is shown in Table 4–1.

The AC to DC SCR units impose a nonsinusoidal current waveform on the AC bus since DC motor torque and speed are a function of the phase-controlled power through the SCRs. This nonsinusoidal, pulsating current appears as a low power factor load to the AC generating system and requires generator kva ratings somewhat larger than for a normal linear load. In most cases, this additional kva rating is insignificant since duty cycle loading on a rig varies considerably during the drilling cycle. Rarely does the system require that all drilling motors operate simultaneously at full torque, low speed—a low power factor area of operation. The best power factor from an SCR unit is achieved at full speed, full load, and is usually over 0.9 under these conditions.

Table 4–1
General Classification of Electric Drilling Rigs by Drilling Depth

System type	Drilling depth (ft)	Drawworks drive HP & total DC mtrs	Mud Pumps drive HP & total DC mtrs	Rotary table drive HP & total DC mtrs	Engine generator system	SCR bays	Generator control
I	0-7500	500 HP 1 mtr	2-500 2 mtrs	Mech coupled to DW	2–500 kw	3	2
II	5000–9000	700 HP 1 mtr	2–500 2 mtrs	Mech coupled to DW	2–800 kw	3	2
III	7000–12,000	1000 HP 1 mtr	2–1000 2 mtrs	Mech coupled to DW	3–600 kw or 2–1050 kw	3	2 or 3
IV	10,000–16,000	1500 HP 2 mtrs	2–1300 4 mtrs	Mech coupled or 400 HP 1 mtr	3–1300 kw	4	3
V	12,000–20,000	2000 HP 2 mtrs	2–1600 4 mtrs	400 HP 1 mtr	4–1300 kw	4	4
VI	15,000–30,000	3000 HP 3 mtrs	3–1600 6 mtrs	400 HP 1 mtr	4–2000 kw or 5–1500 kw	6	4 or 5

Fig. 4–1 Land rig SCR/AC assemblies (3 bays) (Courtesy Baylor Co.)

Most SCR drive manufacturers protect the AC bus by limiting the motor acceleration rates and current to acceptable levels to the system. Often, the driller will depress the drawworks foot throttle to its limit in accelerating the traveling block. Without current limiting, the current through the motor may damage the commutator and trip the AC circuit breaker to the SCR.

Proper selection of the AC circuit breaker supplying each SCR with AC power prevents a complete system shutdown in the event of malfunction of one SCR/motor drive. The trip and interrupting ratings of the SCR circuit breaker must be coordinated with the generator circuit breakers on line (Figs. 4–1, 4–2).

SCR Variable-Frequency Drives

Used in industrial and commercial applications for many years because of their reliability and low maintenance. AC drives employing variable frequency will one day find application in the drill rig and well servicing market. These controls operate directly from the system AC bus and convert the fixed voltage and frequency into variable voltage and frequency to power AC motors at variable speed.

Fig. 4–2 Offshore rig AC/DC SCR assemblies (4 bays) (Courtesy Baylor Co.)

Since the three-phase AC induction motor has no brushes or commutator, it is highly reliable, has low maintenance, and offers the drilling operator superior operating features. Once technology introduces frequency converters at acceptable cost levels, these AC units will be applied to drill rigs and thruster applications.

Some of the features offered by the variable-frequency drive system include the ability to provide full motor torque capability to 100% and beyond over the full-speed range, ability to regenerate when speed deceleration is required, and electronic reversing that eliminates the use of electromechanical contactors (Fig. 4–3).

The All-AC Rig™

The use of AC induction motors coupled to AC variable frequency controls provides an interesting concept. If the diesel/AC generators were used to produce 480 volts alternating current (VAC), three-phase, 60 Hz power to the

---

*The All-AC Rig is a registered trademark of Electrie Drilling Systems, Inc., Houston, TX.

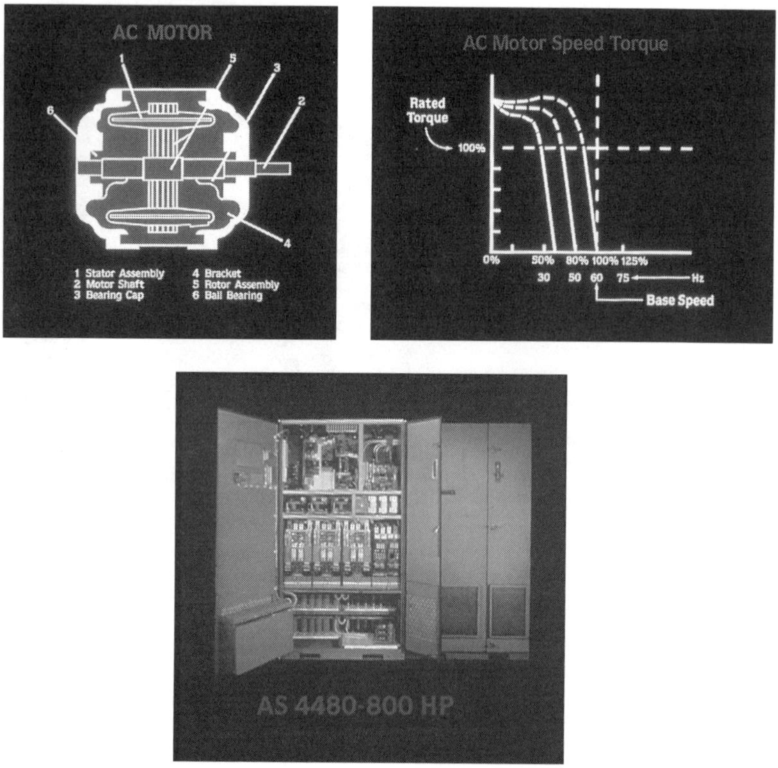

Fig. 4–3 Variable-frequency induction motor drive system (Courtesy Electric Drilling Systems, Inc.)

system bus, a system using all AC motors and components could be designed for maximum efficiency and lowest maintenance and repair. The system would eliminate the use of the 600/480 volt, three-phase transformer normally used for the motor control center and 460-volt motors.

The All-AC Rig™ concept uses the following equipment to provide the electrical and mechanical power for rig functions:

AC synchronous generators, rated 480 VAC, three phase, 60 Hz.

AC induction motors, rated 460 VAC, three phase, 60 Hz for the fixed speed rig applications.

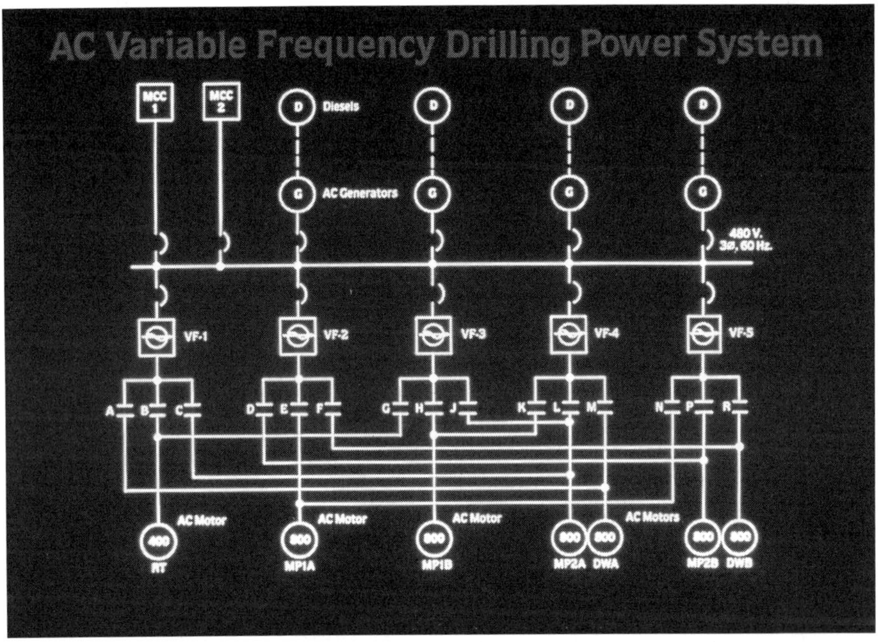

Fig. 4–4 The All-AC Rig™ one-line diagram (Courtesy Electric Drilling Systems, Inc.)

AC variable frequency converters, rated 480 VAC, three phase, 60 Hz, to 0–460 VAC, three phase, 0–70 Hz.

AC induction motors, rated for variable speed operation at a voltage range of 0–460 VAC, three phase, 0–70 Hz.

AC driller's console, rated to control the variable-frequency drive system.

AC lighting transformer, rated 480/120–208 VAC, three phase, 60 Hz.

A typical one line diagram of the All-AC Rig™ is shown in Figure 4–4.

Electric Brakes

The eddy current electric brake has been used as an auxiliary drawworks speed retarder to the primary mechanical brake. The brake control derives its power from the AC bus, stepped-down through a transformer and converted to variable DC through a solid-state power converter. This variable DC controls the brake torque by creating eddy current losses in the rotor from the electromagnetic stator. The greater the DC coil current, the stronger the brake torque developed (Fig. 4–5).

A circuit breaker provides the disconnecting means to the brake control system in the event an electrical fault occurs or shutdown is required.

Thrusters and Propulsion Units

While main propulsion power on drilling vessels is derived directly from dedicated diesel engines in some cases, variable-speed power from an SCR drive is another alternative. Greater control of speed and torque to the propellers can be achieved with the SCR system, including constant horsepower control, thrust limiting, reversing, and regeneration for fast stopping or overhauling conditions.

Using SCR-controlled power directly from the common AC system bus allows more efficient use of engine-generators, since propulsion and drilling functions rarely operate simultaneously. The SCR unit that operates one of the thrusters can be switched over automatically to operate one of the drilling functions and allows each of the control equipment to serve multiple functions on an offshore rig.

The torque developed by a thruster is approximately proportional to the square of the propeller speed and results in a horsepower delivery requirement that varies with the cubic (third power) of the propeller speed. The SCR system and power required from the AC system should be rated based on the maximum anticipated speed (Fig. 4–6).

In some cases, azimuthing or rotating a thruster to change thrust direction is required, and small horsepower fixed- or variable-speed drives are used. Accurate positioning servos can be achieved with a reversing SCR control and position feedback transducer.

Drawworks

The drawworks on a drill rig provides the power for lowering the drillpipe or casing into the hole or removing the drillpipe for bit changes and is a key element on the rig. Drawworks vary in sizes from a single-motor unit for drill depths down to 10,000 ft to a three-motor drawworks for drilling down to 30,000 ft.

Variable speed and reversing under varying loads are the key requirements

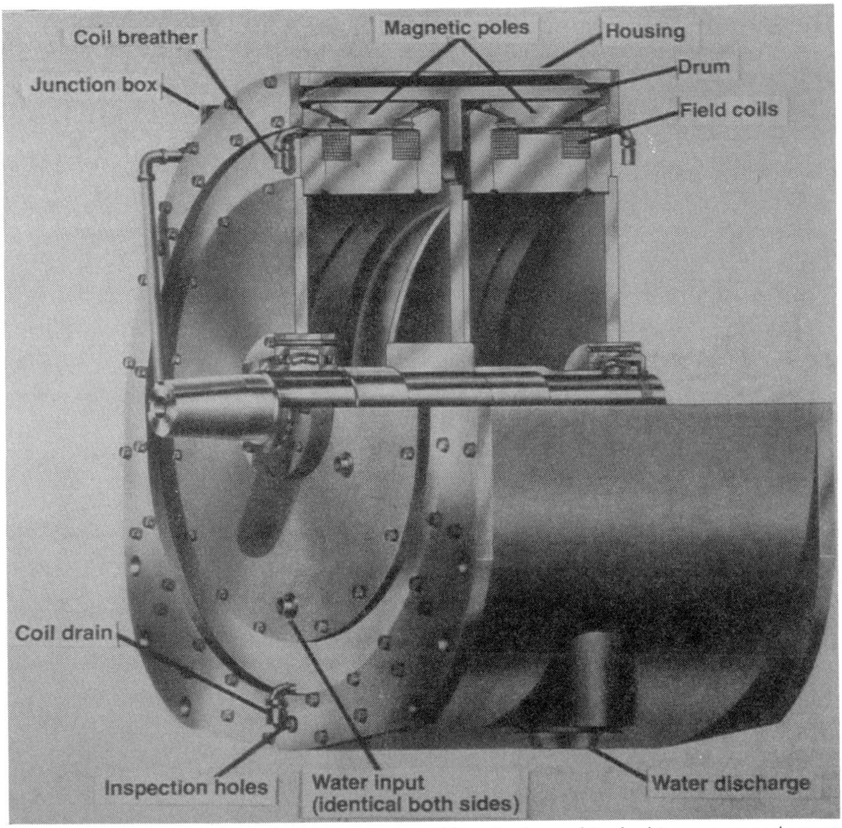

Fig. 4-5. Sectional view of eddy current auxiliary brake and typical torque-speed curves for four coil current levels (Courtesy Baylor Co.)

of the drawworks. Complete control of the drillstring by the driller is essential for safe, efficient operation of the drilling operation; SCR drives achieve speed control, reversing, and dynamic braking of the drawworks.

Rotary Table Drives

Rotary tables can be powered independently, such as on larger deep-well systems, or driven from the drawworks through a mechanical coupling on small rigs. Horsepower requirements for these units are typically less than the

Fig. 4–6 Thruster for offshore drilling vessel (Courtesy Baylor Co.)

drawworks horsepower, but variable speed and reversing are required. Achieving these functions on an electric drilling rig is usually from an SCR system.

Mud Pump Drives

Power for the drilling mud circulating system is obtained from the mud pumps, which are usually driven by a variable-speed DC motor. The mud pumps are positive displacement types and the circulating types and the circulating flow rate is proportional to the mud pump speed. SCR controls with DC motors are usually the primary means for speed control of these pumps which may have one or two motors, depending on the well depth and mud pump size.

Jacking, Winch, and Crane Drives

Many nondrilling functions found on offshore rigs use electrical power. The jacking system on jackups uses constant-speed AC motors directly or as power for hydraulic jacking units. In some cases, SCR controls operate DC motors in the jacking process.

Winches for anchor chains or cables often use electric SCR drives for power control. Variable speed and reversing are required for such applications (Fig. 4–7). Cranes use variable-speed electric drives for hoisting and lowering a load and for rotating and, possibly, advancing the cab along a track.

Automatic Controls and Other Electronic Loads

Electric power is being used more on drilling systems, primarily offshore, for equipment which monitors, measures, and controls drilling and rig activity. Computers, data-logging equipment, vessel controllers, undersea television, dynamic positioning equipment, and communication gear require electrical power for operation at a low-power consumption level. Also, work areas, labs, test equipment, and maintenance facilities require electrical power for their activities.

Instrumentation

Floating drilling vessels principally use electrical instruments for wave height measurement, vessel heave, wind velocity and direction, temperature, and directional drilling measurement. AC power is universally used for such instruments for proper operation.

The information that instruments provide on a drilling rig include:

Drawworks:
- Traveling block position
- Line speed
- Hook weight
- Deadline weight
- Weight on bit

Mud Pumps:
- Pump speed
- Total pump strokes
- Mud flow rate
- Pump pressure
- Riser angle

Rotary Table/Top Drive:
- Rotary table speed
- Rotary torque
- Tong torque

Mud Pits:
- Mud pit volume
- Temperature
- Density
- Gas presence

Floating vessel data:
- Anchor line tension
- Heave rate
- Heave magnitude
- Riser tension
- Crane loads

These data are gathered for economic considerations and for safety to personnel and equipment. Each instrument provides data to improve drilling performance, reduce equipment wear, anticipate well gas and potential blowouts, and prevent overloads. Each rig has all or some combination of electrically operated instruments to record rig data.

On most rigs, the instruments are individually mounted and terminated to sensors located at the point of measurement. Today, sensor data are combined into a single readout console, or data acquisition logger, to reduce the number of readouts, space, and weight. In addition to readout displays, the data can be chart-recorded for display or computer-disk recorded for future analysis and record (Fig. 4–8).

Fig. 4–7 Anchor chain winch with electric drive (Courtesy Baylor Co.)

Analyzing Load Kva, Kw, Kvar, and Power Factor

Every electrical load placed on the drilling rig system consumes electrical power as long as it is performing useful work. Since the rig is a power system unto itself with its own generating equipment and loads, an understanding of the various power components is useful.

Prime mover/AC generator ratings

The prime mover used for the power generating system is the diesel engine or turbine. This device uses gas or fuel oil as the source of energy to create torque and speed at its shaft and contains sufficient BTUs to produce the necessary kilowatt-hours of energy in the conversion process. Since the prime

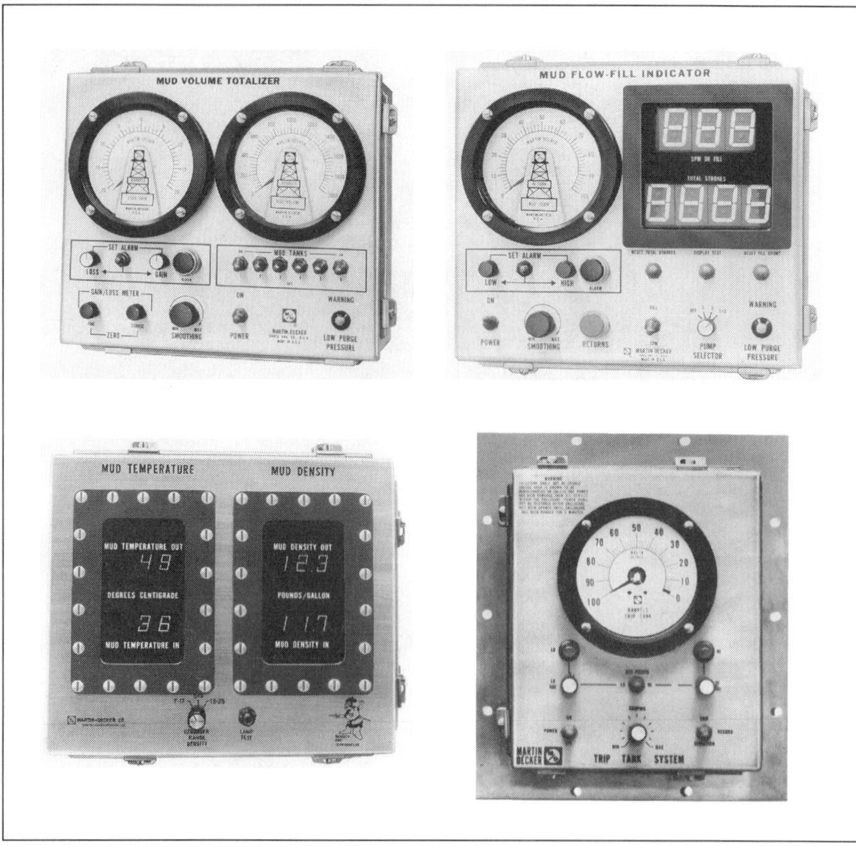

Fig. 4–8 Drilling mud instruments (Courtesy Martin-Decker Co.)

mover develops torque and speed at its shaft, it is a kilowatt-producing machine since:

$$HP = \frac{(T \times N)}{5252},$$

with T = torque in ft-lb
N = speed in rpm

and

$$kw = \frac{HP}{.746}$$

On the other hand, the AC generator is an electrical energy-producing device that produces voltage and current at a certain frequency and phase at its terminals. For electric drilling rigs, the standard voltage for the three-phase generator output is 600 VAC, three phase, 60 Hz.

Therefore, the generator develops *apparent power,* or kilovolt-amperes at its terminals.

For a three-phase generator, its kva rating is:

$$\text{kva} = \frac{(\text{V x I x } 1.732)}{1000}, \text{ kilovolt-amperes}$$

where V = Line-Line three-phase voltage, volts
I = Full-load line current, amperes
1.732 = a constant, the square root of 3
1000 = a constant, to convert volt-amperes to kilovolt-amperes

Since electrical loads consist of power-dissipating components, such as resistors, as well as power storage components, such as inductors and capacitors, kva can be broken down into its components of *real* and *reactive power* or kw and kvars, respectively.

kw = real power component of power, or useful power in kilowatts
kvars = reactive power component of power, or useless power in kilovars

Kva, kw, and kvars can be related by a vector diagram since kw and kvars are in quadrature from each other and kva is the resultant when vector addition is used. In graphic form, a right triangle best illustrates these relationships:

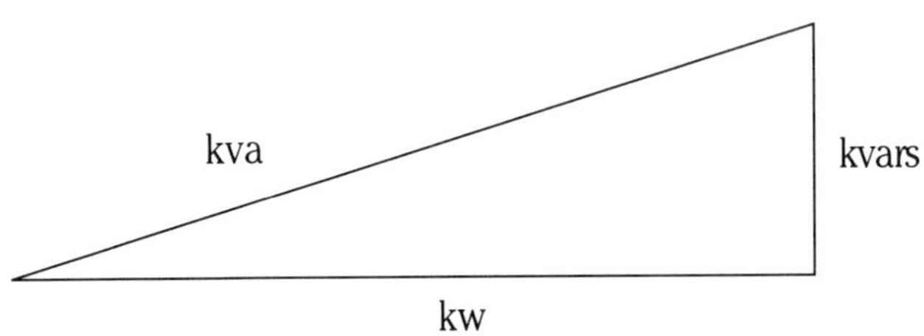

Using the Pythagorean method of calculating the sides of a right triangle, the following equations result:

$$\text{kva} = \sqrt{\text{kw}^2 + \text{kvars}^2}$$

and $$\text{kw} = \sqrt{\text{kva}^2 - \text{kvars}^2}$$

and $$\text{kvars} = \sqrt{\text{kva}^2 - \text{kw}^2}$$

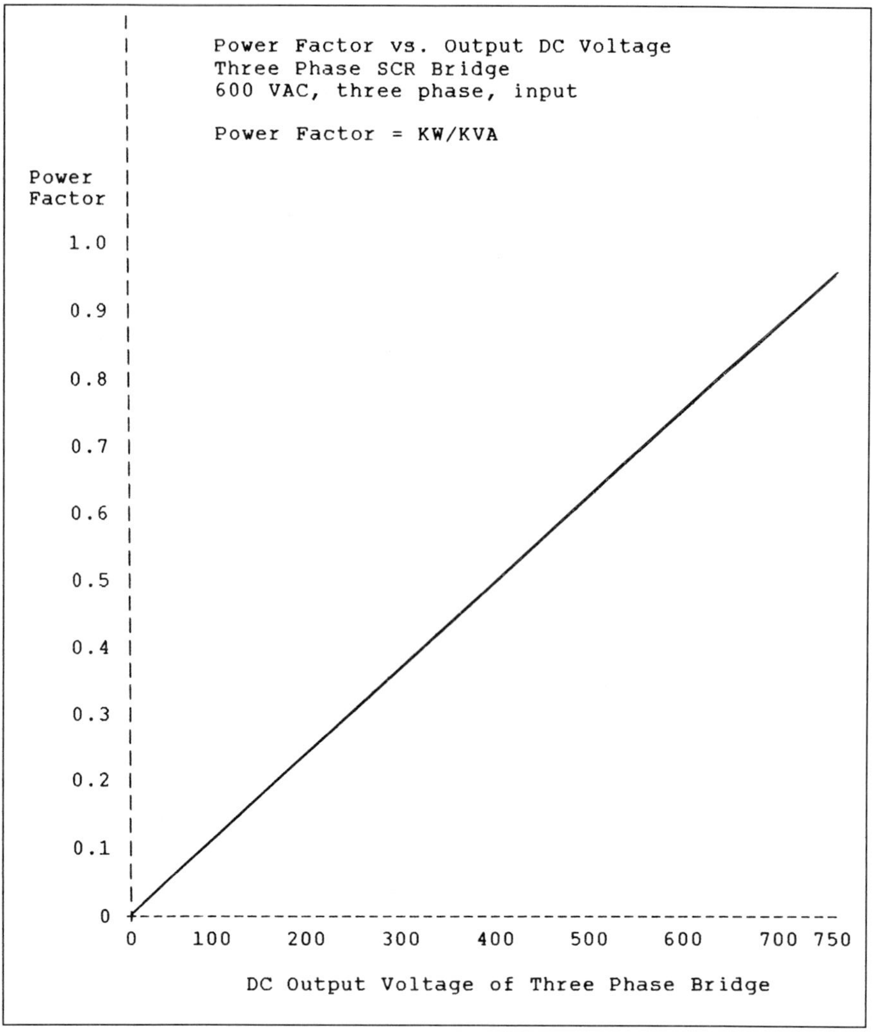

Fig. 4–9 Power factor produced by a variable voltage, three phase, full wave, SCR bridge

Therefore, an AC generator can deliver kva, kw, and kvars to its loads. Its maximum kw capability is that produced by the prime mover (assuming the generator is capable of this rating), while its kva capability is determined by the physical design of the generator itself.

While impractical, the desirable load is one where no reactive load, or kvar, exists and where kw equals the kva. Since all loads have a form of inductance or capacitance, and SCR systems introduce an *apparent lagging current,* kvars exist in a real and practical sense on drilling rigs. In many cases, the kvars can represent a sizable portion of the AC load to the generator. In terms of generator ratings, the kvars vectorally added to the kw of the load represents the kva requirements on the generator and thus its size, weight, and cost.

A measure of the relative real power to the apparent power is the *power factor* (PF) of the load. It can be measured in terms of the ratio of the real component of current to the total current or the ratio of the real power to apparent power. In sinusoidal terms, power factor is the cosine of the angle between the applied voltage and the resulting current in the load. However, for reasons of simplicity, this is best expressed in terms of a ratio of real power to apparent power. In equation form, this is:

$$PF = \frac{kw}{kva}$$

Since kw is always less than or equal to kva, the ratio of PF is always 1.0 or less. Examples of various loads and their power factor are given below:

AC induction motors: PF = 0.8
Resistive heaters: PF = 1.0
Incandescent lamps: PF = 1.0
SCR, three-phase bridges: PF = variable, depending on the output voltage (Fig. 4–9).

Bibliography

Terbeck, William, and George S. Stivers. "Drilling Rig Instruments Promote Efficiency, Safety." *Petroleum Engineer International,* May 1979.

5 THE SILICON-CONTROLLED RECTIFIER (SCR)

Few power control devices have had the impact of the silicon-controlled rectifier (SCR). This semiconductor, developed in 1957 by the General Electric Company, is a member of the growing family of controlled switching devices known as thyristors. The thyristor group includes SCRs, triacs, and gate turn-off devices. All are characterized by their ability to switch electrical power statically via a controlled, low-power gate signal (Fig. 5–1).

An SCR is like a rectifier; i.e., it blocks power in its reverse direction and allows power to conduct in the forward direction. The SCR is a unique rectifier with a gate conductor that determines when power may conduct in the forward direction. Current ratings of such devices range from several hundred milliamps (ma) to several thousand amps with blocking voltages up to several thousand volts.

SCRs are used in many applications and industries as various forms of power control elements. The following list indicates the varied uses of the SCR:

Military and Aerospace:

- Frequency converters
- DC to AC inverters
- AC power supplies
- DC power supplies
- Static switches
- Voltage regulators
- Static exciters

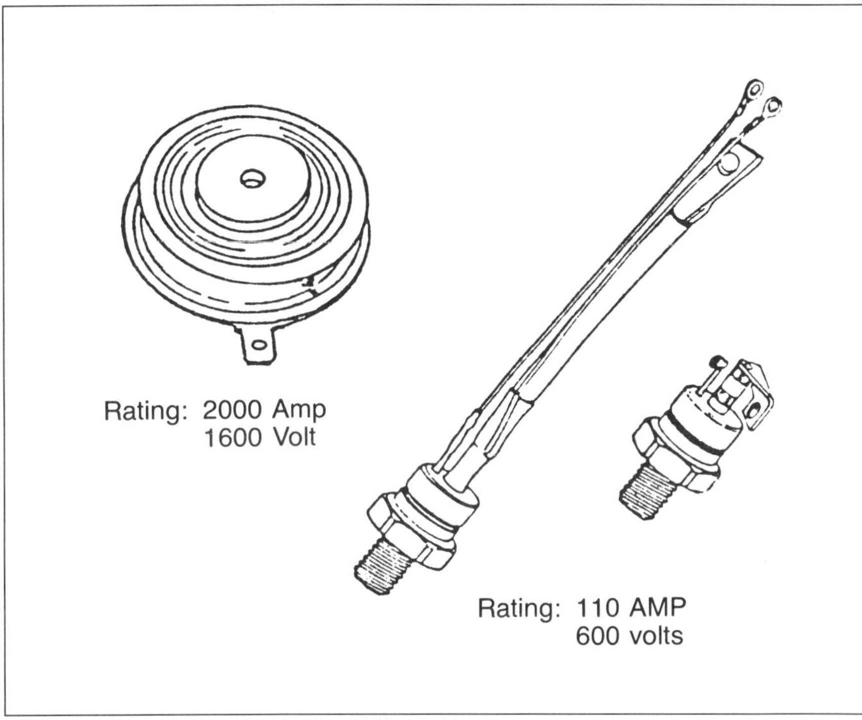

Fig. 5–1 Power SCRs used in the drilling industry

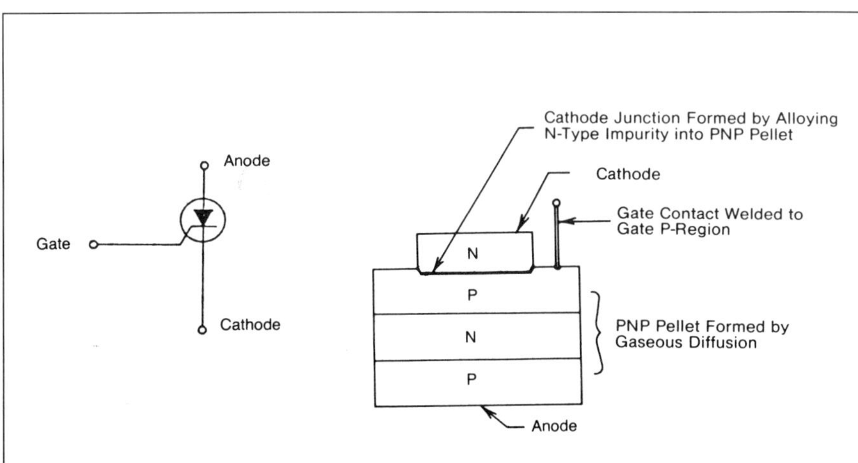

Fig. 5–2 SCR symbol and structure

Computers:

- •Uninterruptible AC power supplies
- •DC power supplies
- •Static AC transfer switches
- •Disk drives
- •Logic switches

Industrial:

- •DC mill and conveyor drives
- •AC variable-frequency drives
- •Boiler controls
- •Knitting and textile drives
- •Extrusion drives
- •Boring, milling, and cutting machines
- •Motor starters

Drilling and Mining:

- •Drawworks, mud pumps, rotary table, and top drives
- •Ore haul trucks
- •Underground power system
- •Downhole pumps
- •Downhole drilling
- •Thruster drives

Petrochemical Processes:

- •Oil/salt water separators
- •Pipeline pump drives
- •Variable-speed pumps

SCR Theory

To understand how the SCR functions, refer to Figure 5–2, showing the schematic diagram of the device and its basic structure.

The schematic symbol of the SCR shows the three terminals of the device: the anode, cathode, and gate. Forward current conduction of the SCR is from its anode to its cathode, and the gate signal flow is from the gate to the cathode. Without gate current, the SCR blocks current flow in the forward direction but always blocks conduction in the reverse direction when operated within its voltage rating. In essence, it performs a rectifying function only when conduction is initiated with gate current at the time the anode is positive with respect to its cathode.

The SCR structure in Figure 5–3 explains the physics action of the device in both its blocking and conductive states. Four layers of alternate P- and N-type semiconductor material constitute an SCR. For power SCRs, the common fabrication process is an alloy-diffused technique performed in an extremely clean environment. This minimizes any impurities to ensure long-term life and stability of the SCR's electrical characteristics. To guarantee this stability under use, each SCR is welded and hermetically sealed after the wafer is fabricated and assembled.

One method which explains the regenerative operation of the SCR is the "two transistor analogy." In Figure 5–4, the P-N-P-N structure is divided into two equivalent transistors.One of the transistors is a PNP-type and the other is an NPN-type. With no control current applied to the control terminal, the pair operate in a forward-blocking "off" state. Initiation to the low-impedance, or conducting, state is started by applying a control current to the NPN transistor. The collector develops current flow and initiates the turn-on of the PNP transistor which enhances the turn-on of the NPN transistor.

This regenerative process occurs rapidly until both transistors are fully conducting. Removing the external control current will not turn off the pair of devices, and the complementary switch remains on until the load current is interrupted. If this interruption of current occurs and no control current is present, the pair again revert to a forward-blocking state.

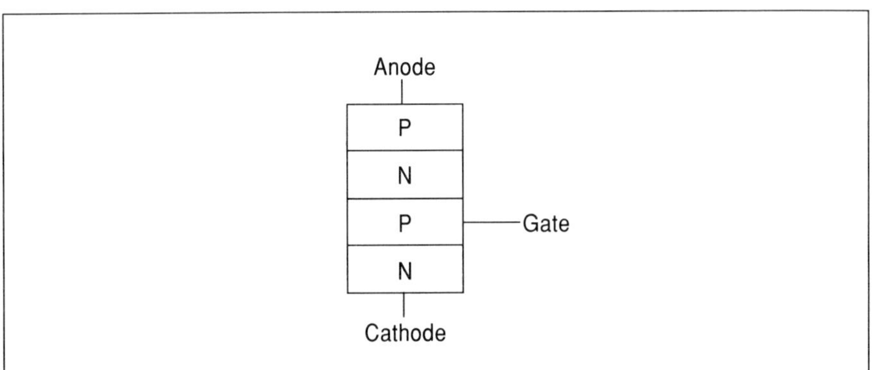

Fig. 5–3 SCR PNPN structure

This is essentially the action of the SCR. Once the SCR is forward-biased, a gate control current is applied and the SCR converts from a blocking condition to a conducting state. Removing the control gate current does not turn the SCR off, and it continues to conduct as long as it remains forward- biased and anode current continues to flow. Only by interrupting this current will the device turn off and block in the forward direction.

Other means can turn on an SCR besides gate current, some of which are undesirable in SCR power circuits. Three of these alternatives are forward voltage breakover, high rate-of-change of voltage, and excessive junction temperature.

Forward Voltage Breakover

Increasing the forward voltage of the SCR increases the forward leakage current to a level where regenerative action occurs within the device. When this occurs, a form of avalanche breakdown occurs at V_{BO}, characterized by a sharp increase in SCR current (Fig. 5–5). Thus, the SCR must be kept within its voltage rating to prevent nuisance operation.

Rate-of-Change of Voltage

Since the P-N junctions of the SCR have a finite amount of capacitance, a step change in voltage across the SCR develops a current flow from one junction to another. The amount of this current is a function of the value of junction

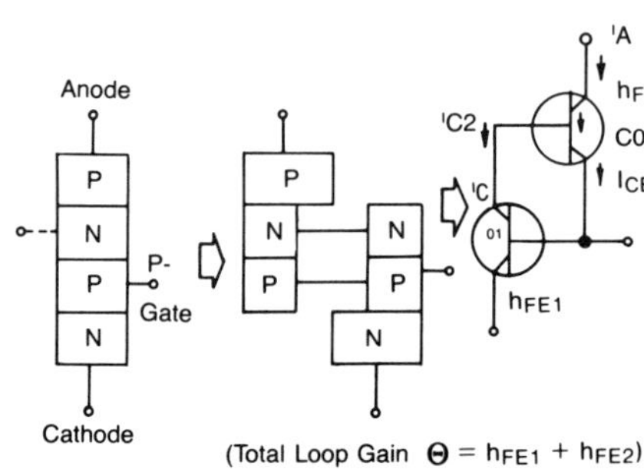

Fig. 5–4 SCR/two transistor analogy

capacitance and the rate of voltage rise across the SCR. Expressed as an equation, the current in amps through the SCR is:

$$I = C\frac{dV}{dt} \text{ amps}$$

where C = Capacitance, farads
dV = Change in voltage, volts
dt = Change in time corresponding to the voltage change, seconds

For a given SCR with junction capacitance C, the current is proportional to the dV/dt. The faster the voltage rate of rise, the greater the current and, at a given current level, the SCR turns on without external gate current being applied.

In the event fast voltage rates exist in an SCR circuit, series resistor-capacitor components across the SCR may limit this rate (Fig. 5–6).

Junction Temperature

Another parameter that causes SCR turn-on without external gate current is excessive device temperature. The leakage current in SCRs doubles for every 8°C increase in temperature. If the current reaches a sufficient level, the SCR will turn on. To prevent this action, SCRs must be kept adequately cooled below their maximum junction temperature.

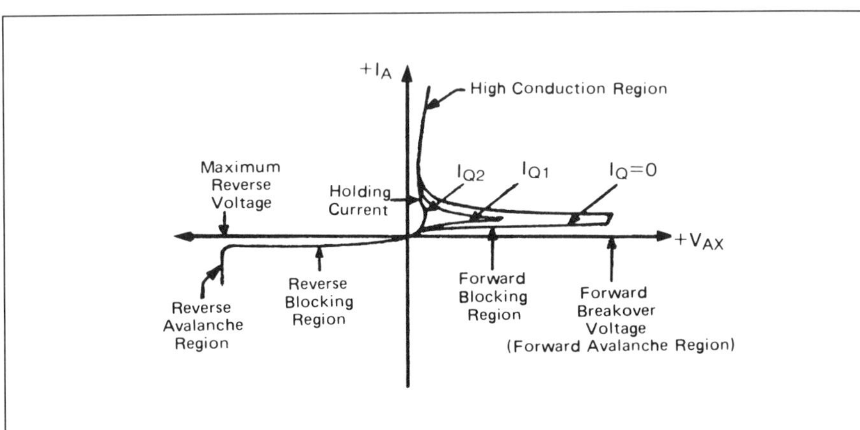

Fig. 5–5 SCR volt-ampere characteristics

SCR Characteristics

Since the SCR is a regenerative device, it can only be turned off by reducing the anode current below its "holding current" level. This process of turn-off is called "commutation" and is accomplished in several ways in circuits.

Line Commutation: Used in AC circuits where the AC line voltage reversal each half-cycle produces a reverse SCR voltage and this commutates, or turns off, the SCR.

Self-Commutation: The SCR converter produces a current into a load that causes the current to go to zero before the desired cycle is over.

Forced Commutation: A separate circuit produces a reverse voltage to extinguish the SCR current.

The power loss in an SCR is the product of its voltage and current. During its off condition, the power loss equals blocking voltage multiplied by its leakage current. In most devices, this power loss is insignificant, amounting to less than 1% of the device power rating. During the SCR conducting period, the SCR dissipates 99% of its losses since the forward load current is high. Typically, the loss in an SCR is:

$$P_{SCR} = (V_F)\,(I_F), \text{ watts}$$

where V_F = RMS forward voltage drop, volts
I_F = RMS forward current, amps

The SCR forward voltage drop is usually 1.5–2.5 volts for most SCRs and depends on the forward current.

In many cases, conduction occurs only for half the time or less; this must

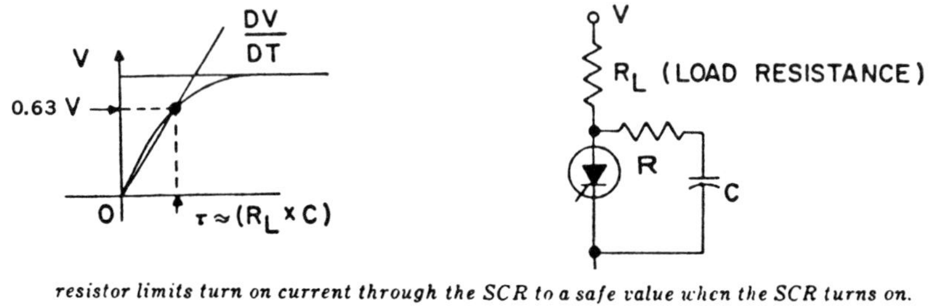

resistor limits turn on current through the SCR to a safe value when the SCR turns on.

Fig. 5–6 dV/dt suppression network across SCR

be considered in calculating power loss. Most SCR manufacturers provide curves of average current vs average power dissipation for various duty cycles and waveforms. These curves help determine power loss under varying conditions and can determine SCR temperature if the thermal resistance is known. A typical curve is shown in Figure 5–7.

Another characteristic of an SCR is its thermal resistance. This resistance determines the SCR's ability to dissipate its internally generated heat, or power loss. A series of thermal resistances exists in the assembly of an SCR, including its metal heat sink and the surrounding cooling medium. To calculate the total thermal resistance of an SCR assembly, the following must be considered:

$\theta_{J\text{-}C}$ = thermal resistance from SCR junction to its case, °C/watt

$\theta_{C\text{-}HS}$ = thermal resistance from SCR case to metal heat sink, °C/watt

$\theta_{HS\text{-}A}$ = thermal resistance from heat sink to ambient air, °C/watt

θ_{T} = $\theta_{J\text{-}C} + \theta_{C\text{-}HS} + \theta_{HS\text{-}A}$ = total thermal resistance of SCR assembly, °C/watt

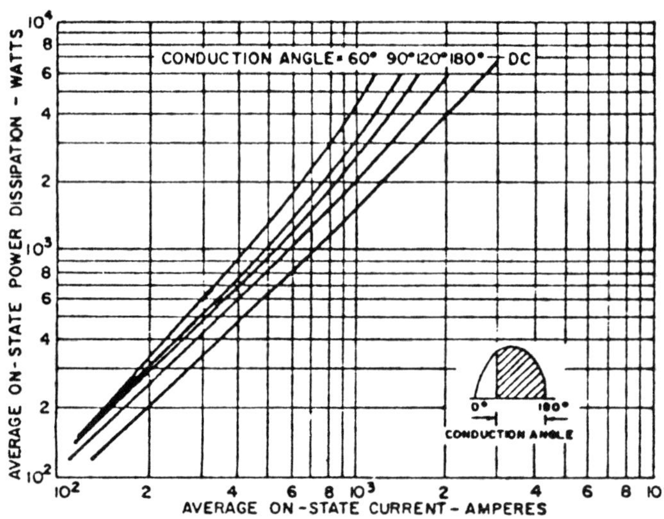

Fig. 5–7 SCR current-power characteristics

Often, the SCR temperature rating is measured in terms of its maximum junction temperature and is usually 125°–200°C. If the thermal resistance (O_T) is known, the temperature of the air leaving the heat sink (T_A) is known. If the power dissipated by the SCR (P_{SCR}) is known, the SCR junction temperature can be calculated by the following formula:

$$T_J = (P_{SCR})\ (O_T) + T_A),\ °C.$$

An SCR can be fused by matching the SCR surge rating to the fuse rating. The figure given by most manufacturers is an I^2t rating for short-duration RMS currents (I) for a given time (t) of less than one-half cycle of the AC power line. Using a fast-acting fuse whose melting I^2t is less than that of the SCR provides fault protection for the SCR.

Providing adequate gate current signals to the SCR is essential in maintaining reliability of both the SCR and the circuit it is used in. The SCR specifications define a range of current values for which the SCR always "fires," or begins conduction, and the maximum limit that may damage the SCR. Between these limits, the circuit designer finds the optimum gate current signal. The most desirable signal has fast rise time (less than 50 nanoseconds), high initial current amplitude (1 ampere for large hockey-puck type SCRs), and a duration which can assure conduction of the SCR into the load.

The following definitions and terms for an SCR are given as an aid in understanding its technology and are approved by IEEE Standards on solid-state devices.

SCR Terminology and Characteristics

Terminology	Symbol	Definition
Repetitive peak reverse voltage, gate open	V_{ROM}(rep))	Maximum allowable instantaneous value of repetitive reverse (negative) voltage that may be applied to the reverse-blocking thyristor anode terminal with gate terminal open. While this value of voltage does not necessarily represent a "breakdown" voltage, it should never be exceeded except by transient rating, if the device has such a rating. Reverse-blocking thyristors can handle this voltage satisfactorily with negative gate voltage also.

Terminology	Symbol	Definition
Nonrepetitive peak reverse voltage, gate open	V_{ROM} (nonrep)	Maximum allowable instantaneous value of reverse (negative) voltage including all non-repetitive transient voltages but excluding all repetitive transient voltages, that may be applied to the reverse-blocking thyristor anode terminal with gate terminal open. Reverse-blocking thyristors can handle voltage satisfactorily with negative gate voltage, also.
Peak forward-block voltage, gate open	V_{FOM}	Maximum instantaneous value of forward-blocking voltage (anode terminal positive), including transient voltages permitted by the manufacturer under stated conditions and which will not switch the reverse-blocking thyristor to the on-state.
Peak forward voltage	PFV	Maximum instantaneous value of forward voltage permitted by the manufacturer under stated condition and which may cause the reverse-blocking thyristor to switch to the on-state. If anode breakover occurs at a voltage lower than the PFV, no damage to the reverse blocking thyristor will occur.
Average forward current, on-state	$I_{F(AV)}$	Maximum continuous DC current which may be permitted to flow in the forward direction (from anode to cathode) under stated conditions of frequency, temperature, reverse voltage, and current waveform.
Forward break-over voltage	$V_{(BR)F}$	Maximum positive voltage on the anode terminal with respect to the cathode terminal for which the small-signal resistance is zero with stated gate termination. This is also the voltage at which a thyristor switches into the conductive state.

Terminology	Symbol	Definition
Instantaneous forward current on-state	i_F	Instantaneous value of anode current flowing into the thyristor in the conducting state.
Instantaneous on-voltage (forward voltage drop)	V_F	Instantaneous voltage drop between anode and cathode terminals during conduction of current from anode to cathode terminals while the device is in the on-state.
Full cycle average on-voltage (forward voltage drop)	$V_{F(AV)}$	On-voltage averaged over one complete cycle with stated forward current flowing in a 60 Hz, single-phase, half-wave rectifier with resistive load and no circuit-triggering angle delay.
Instantaneous forward gate current	i_{GF}	Instantaneous current flowing between gate and cathode terminals in a direction to forward bias the gate junction.
Instantaneous forward gate voltage	V_{GF}	Instantaneous forward voltage between gate and cathode terminals with anode terminal open.
DC gate trigger current	I_{GT}	Forward gate current required to trigger a thyristor at stated temperature conditions.
DC gate trigger voltage	V_{GT}	Gate voltage with I_{GT} flowing but prior to start of anode conduction.
Effective irradiance to trigger	H_{ET}	The amount of incident radiant flux density which is effective in causing a light-activated device to switch to the conducting state. This is the integral of the product of the spectral response curve of the cell and the spectral distribution of the energy source, expressed in watts per square centimeter, which causes the device to switch.
SCR current	Io	$I_{F(AV)}$ under specific condition of half-wave rectified sine-wave current, 180° conduction angle.

Terminology	Symbol	Definition
RMS forward current, on-state	I_f	Maximum continuous RMS current which may flow in the forward direction under stated conditions. Average forward current rating applies simultaneously.
Peak one-cycle surge forward current	$I_{FM(surge)}$	Maximum allowable nonrecurrent peak current of a single forward cycle (8.3 milliseconds duration) in a 60-Hz, single-phase, resistive load system. The surge may be preceded and followed by maximum rated voltage, current, and junction temperature conditions; maximum allowable gate power may be concurrently dissipated. However, specified limitations on anode current during switching should not be exceeded.
I squared t	I^2t	Measure of maximum forward nonrecurring overcurrent capability for very short pulse durations (8.3 milliseconds or less, unless otherwise specified). I is in RMS amps, and t is pulse duration in seconds. The same conditions as listed above for $I_{FM(surge)}$ apply.
Peak reverse gate voltage	V_{GRM}	Maximum allowable peak reverse voltage between the gate terminal and the cathode terminal.
Peak gate power dissipation	P_{GM}	Maximum instantaneous value of gate power dissipation.
Average gate power dissipation	$P_{G(AV)}$	Maximum allowable value of gate power dissipation averaged over a full cycle.
Holding current	I_H	Value of i_F below which thyristor returns to forward-blocking state after being in forward conduction under stated temperature and gate termination conditions.
Latching current	I_L	Value of minimum anode current for thyristor to switch to the on-state, and to remain in the on-state after removal of the gate trigger pulse under specified conditions.

Terminology	Symbol	Definition
Instantaneous forward-blocking current	i_F	Instantaneous anode current at stated conditions of forward-blocking voltage, junction temperature, and gate termination.
Instantaneous reverse-blocking current	i_R	Instantaneous anode current at stated conditions of negative anode voltage, junction temperature, and gate termination.
Delay time	t_d	Time interval between the time the gate current pulse reaches 10% of its final value and the time when the resulting forward current reaches 10% of its maximum value during switching from the off-state to the on-state into a resistive load under stated conditions.
Rise time	t_r	Time interval between the time the forward current reaches 10% of its maximum value and the time the forward current reaches 90% of its maximum value during switching from the off-state to the on-state into a resistive load under stated conditions.
Turn-on time	t_{on}	Sum of delay time and rise time.
Turn-off time	t_{off}	The time interval between zero current and the time of reapplication of positive forward-blocking voltage under specified conditions with the device remaining in the off-state after being in the on-state.
Reverse recovery time	t_{rr}	The time interval between zero current and the time at which the reverse current through the device has reached a specified value (usually 10% of the peak reverse recovery current) under specified conditions after having been in the on-state.
Thermal resistance	θ	Temperature rise per unit power dissipation of a designated point above the temperature of a stated external reference point under conditions of thermal equilibrium.

Terminology	Symbol	Definition
Transient thermal resistance	$\theta_{(t)}$	Temperature rise per unit power dissipation of junction above stated reference point for specified period of time after application of step increase of junction power dissipation with device case or ambient temperature held constant.
Junction temperature	T_J	Virtual junction temperature
Case temperature	T_C	Case temperature
Ambient temperature	T_A	Ambient temperature
Storage temperature	T_{stg}	Storage temperature

SCR Applications

Since the SCR is a static switch, it does not perform any wave-shaping or analog functions. It either blocks current flow or conducts current in the forward direction at a voltage drop of 1.5–2.5 volts. When the voltage reverses across the SCR, it blocks this voltage within its rating. Thus, the following sample circuits show the SCR as a rectifier whose time for conduction is controlled.

Half-Wave Circuit

When it is desired to operate a load from a half-wave rectifier and control this power, the circuit of Figure 5–8 is used. The main power source is a single-phase AC, and the circuit consists of an SCR with firing circuit and the load.

The voltage waveforms show events occurring in time sequence with a firing pulse of approximately 90°. If the firing pulse were delayed in time during the complete positive half-cycle, the load voltage would be zero. If it were advanced to the beginning of the positive half-cycle, the load voltage would be a complete half-wave. The voltage could never be greater than full conduction of the positive half-wave cycle since the SCR is a unidirectional device (a rectifier) and becomes reverse-biased during the negative half-cycle.

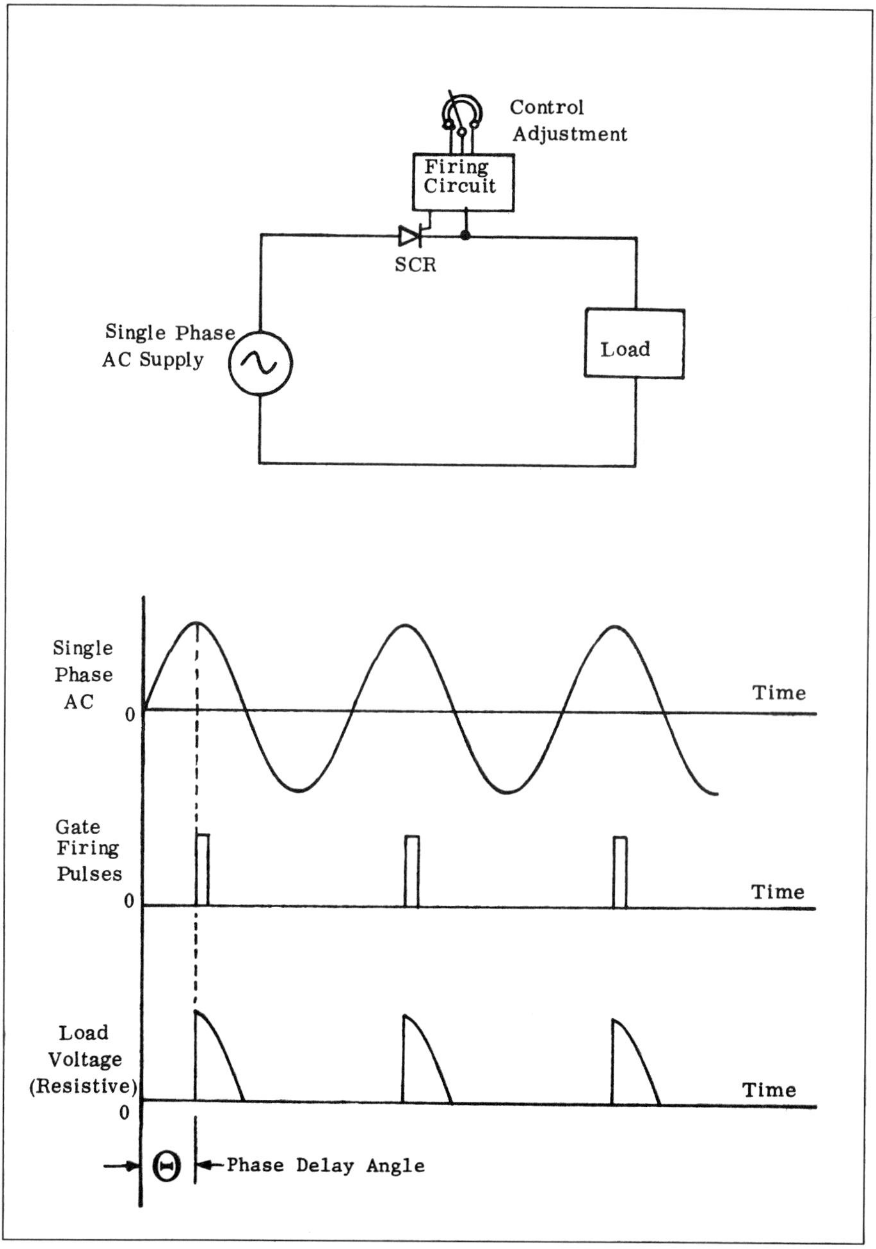

Fig. 5–8 Half-wave SCR circuit and waveforms

When the SCR is full on and producing complete half-wave voltage pulses, the DC voltage to the load is:

$$V_{DC} = \frac{V_{peak}}{\pi} \text{ volts DC}$$

where V_{peak} = peak of the AC line voltage
π = 3.14

A more general equation to include SCR phase-delay angle is:

$$V_{DC} = \frac{V_{peak}}{2\pi}(1 + \cos\theta), \text{ volts DC}$$

The applications for the circuit in Figure 5–8 include simple lamp dimmers, electric tools, or fractional horsepower DC motor controls.

Full-Wave Circuit

Adding four SCRs in a bridge configuration allows full control of the single-phase AC voltage to a DC load (Fig. 5–9).

During the positive half-cycle of the AC source, SCRs 1 and 2 are both forward-biased and in series with the load. When the two SCRs are fired with a gate voltage, they allow conduction of current I_1 into the load for the remainder of the positive half-cycle. During the negative half-cycle, SCRs 3 and 4 are forward-biased and will conduct current into the load when fired. For each case of half-cycle operation, the conducting SCRs are turned off as the voltage goes through zero and reverses polarity.

For this example, the firing angle was set for approximately 90°. If the firing angles were delayed for a full 180°, the output voltage would be zero. Conversely, if they were advanced to fire the SCRs at the beginning of each half-cycle, the output to the load would be a full rectified DC. The value of the full-wave rectified DC with phase delay angle θ for a resistive load is:

$$V_{DC} = \frac{V_{peak}}{2\pi}(1 + \cos\theta), \text{ volts DC}$$

where V_{peak} = peak value of AC line voltage
π = 3.14

Applications of this circuit are in medium-horsepower DC drives and heater controls.

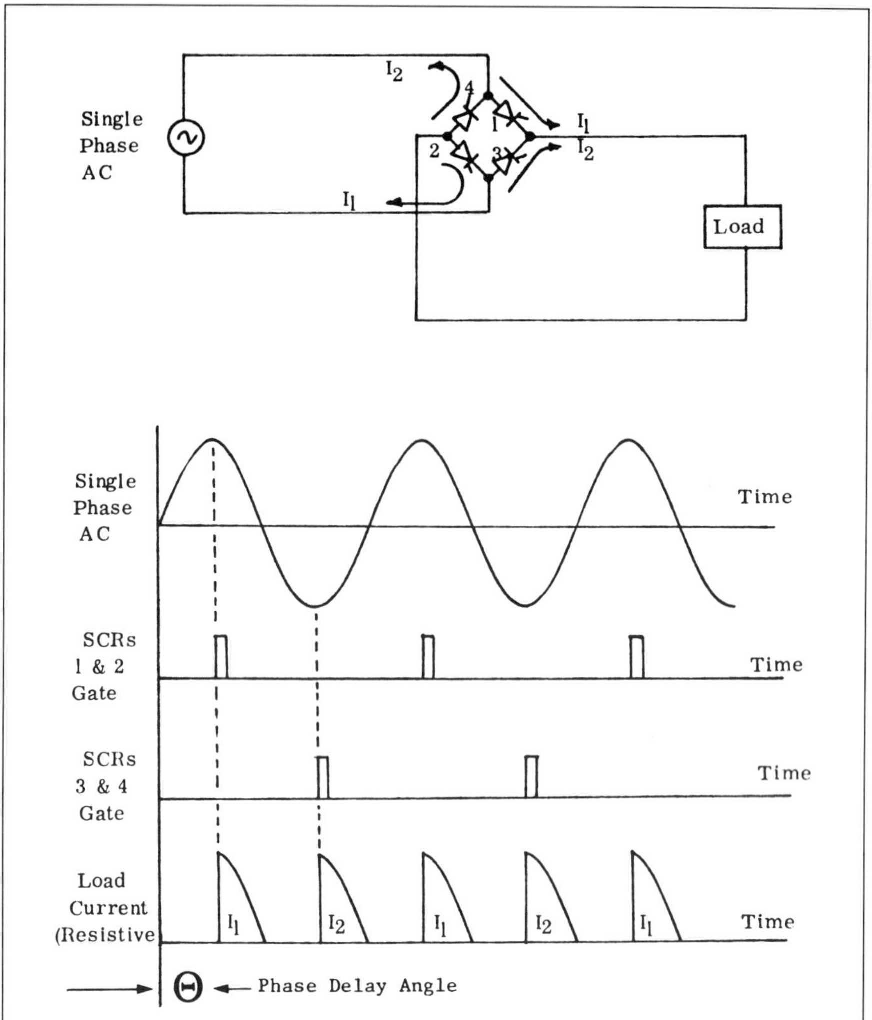

Fig. 5–9 Full-wave single-phase SCR circuit and waveforms

Full-Wave Three-Phase Circuits

If the AC source is three-phase and the load power is controlled with SCRs, the three-phase bridge circuit is used. Since the phase voltages of the AC line are phase-displaced 120° apart in time, each SCR must be fired in proper

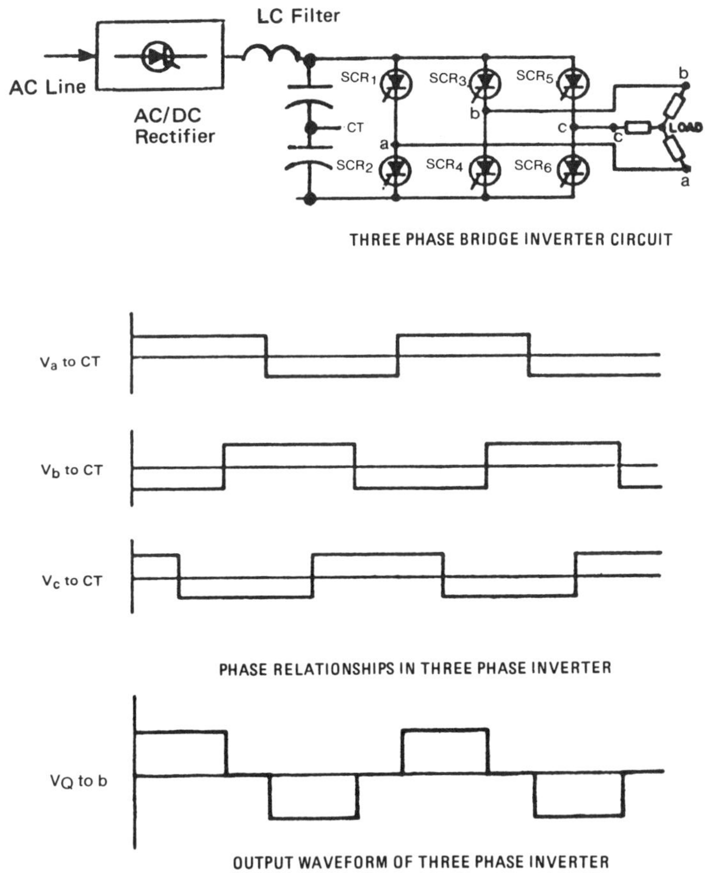

Fig. 5–10 AC/DC/AC SCR power converter with simplified schematic.

sequence to ensure a balanced load voltage. When rectification occurs, the resulting DC consists of pulses whose frequency rate is six times that of the line supply. Delaying the firing pulse to the end of the cycle results in zero output; advancing the firing pulse to the end of the cycle results in zero output; advancing the firing pulse to at least 120° results in full, three-phase rectified output. The equation for the DC voltage when the SCRs are fully rectifying is:

$$V_{dc} = 0.95 \times V_{peak}, \text{ volts DC}$$

AC to DC to AC Converters

In many applications the fixed frequency AC line voltage is converted to another frequency, or variable frequency, for load use. In such cases, SCRs provide multiple power conversions to achieve AC to the load (Fig. 5–10).

The input SCR rectifier converts the input AC to variable DC, filtered by inductor L and capacitor C. In the inverter section, SCR pairs 1 and 2 alternately switch off to allow the phase to go from $+E_{DC}$ to $-E_{DC}$. The same occurs for SCR pairs 3–4 and 5–6. In addition, each pair is switched 120° in time-phase from each other. The result is a three-phase AC stepped-wave voltage at Phase 1', Phase 2', and Phase 3' terminals.

Firing Circuits

SCR control is accomplished by low-power electronics coupled to the SCR through its gate circuit. The control and firing circuit must perform several functions to be useful:

1. The control input is either a manual adjustment by the operator or a signal input from a process signal.
2. The firing angle control must be smooth and continuous.
3. The firing angle control must be continuously synchronized with the AC line each half-cycle.
4. The gating pulse must have a steep rise time of less than 0.5 microsecond, have adequate current delivered to the gate, and be of sufficient duration to satisfy SCR holding current requirements and load requirements.

6 SCR DC AND AC DRIVES

SCR DC Drives

Single-Phase, Half-Wave DC

Early applications of SCRs to the control of DC motor speed and torque were in the area of half-wave drives. A single SCR controlled the power to a series (universal) or shunt field motor. A typical circuit configuration of each is shown in Figure 6–1.

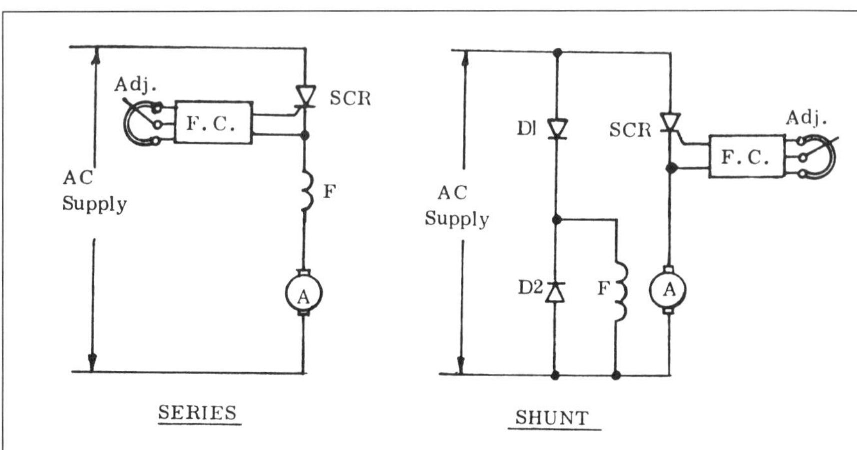

Fig. 6–1 Typical half-wave DC motor drive circuits

Power for the half-wave configuration was received from a single-phase AC supply, and the firing circuits (F.C.) were simple. Power was delivered to the armature only during the positive half-cycles and was phase-controlled by the SCR to adjust the desired speed and torque.

The shunt motor required a separate field supply as shown with diodes D1 and D2. As the SCR firing pulse was advanced, the motor voltage and thus its speed increased proportionally. The current delivered to the armature was

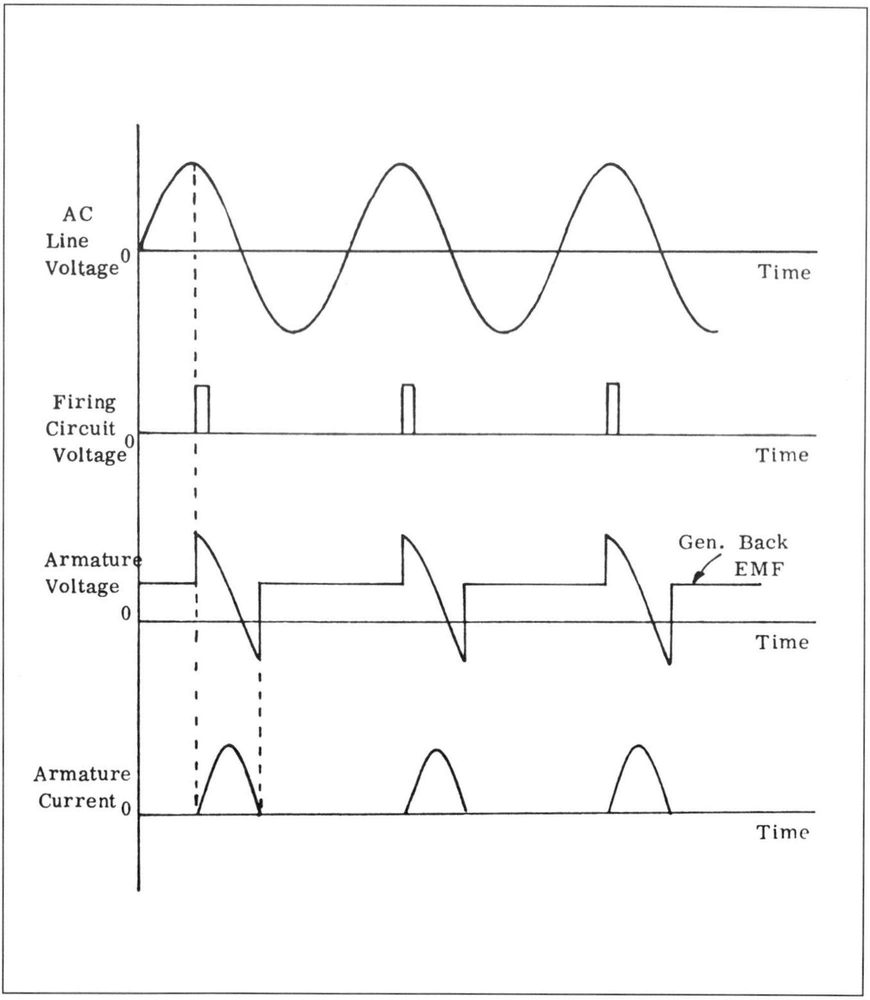

Fig. 6–2 Half-wave SCR motor drive waveforms for a shunt motor.

approximately proportional to the torque demanded by the motor shaft load. During the off or negative half-cycles of the AC supply, the SCR would be reverse-biased by the line voltage and the counter back-emf. Typical waveforms are shown in Figure 6–2.

Single-Phase, Full-Wave DC Drive

Full-wave rectification with SCRs improves speed range and reduces motor heating. SCRs 1 and 2 must be fired together during the positive half-cycle of the AC and SCRs 3 and 4 must be fired simultaneously during the negative half-cycle of the AC. The SCR "bridge" converts all half-cycles to DC at the armature and controls motor speed and torque through the adjustment of the SCR firing angle. An example of firing all SCRs at 90° is shown in Figure 6–3.

Three-Phase, SCR/Rectifier, DC Drive

For larger horsepower systems, the typical AC power source is three-phase. To convert this AC to rectified DC, at least six rectifying elements are required. To produce an economical drive system, some manufacturers use three SCRs and three power rectifiers in a three-phase bridge configuration (Fig. 6–4), sometimes called a "half-controlled bridge."

SCRs 1, 2, and 3 act as the power control elements; rectifiers 4, 5, and 6 provide a return path for current flow. Properly gating the SCRs allows full control of the motor DC voltage from zero to full-wave rectified line voltage. The disadvantage of this system is the asymmetrical currents imposed on the AC line which, in some cases, are undesirable.

Three-Phase, SCR DC Drive

The most popular drive currently used on a drilling rig is the full-wave SCR control operating from a three-phase supply. Its most attractive feature is its relatively low cost since it uses the standard diesel locomotive DC traction motor as part of the drive (Figs. 6–5, 6–6).

Figure 6–7 shows schematic diagrams of SCR-type series and shunt-type motor drives operating from three-phase voltages. Each of the six control elements is an SCR and provides smooth, uniform voltage control to the motor. Each also produces balanced currents on the AC line for better power utilization from the AC generators (Fig. 6–8).

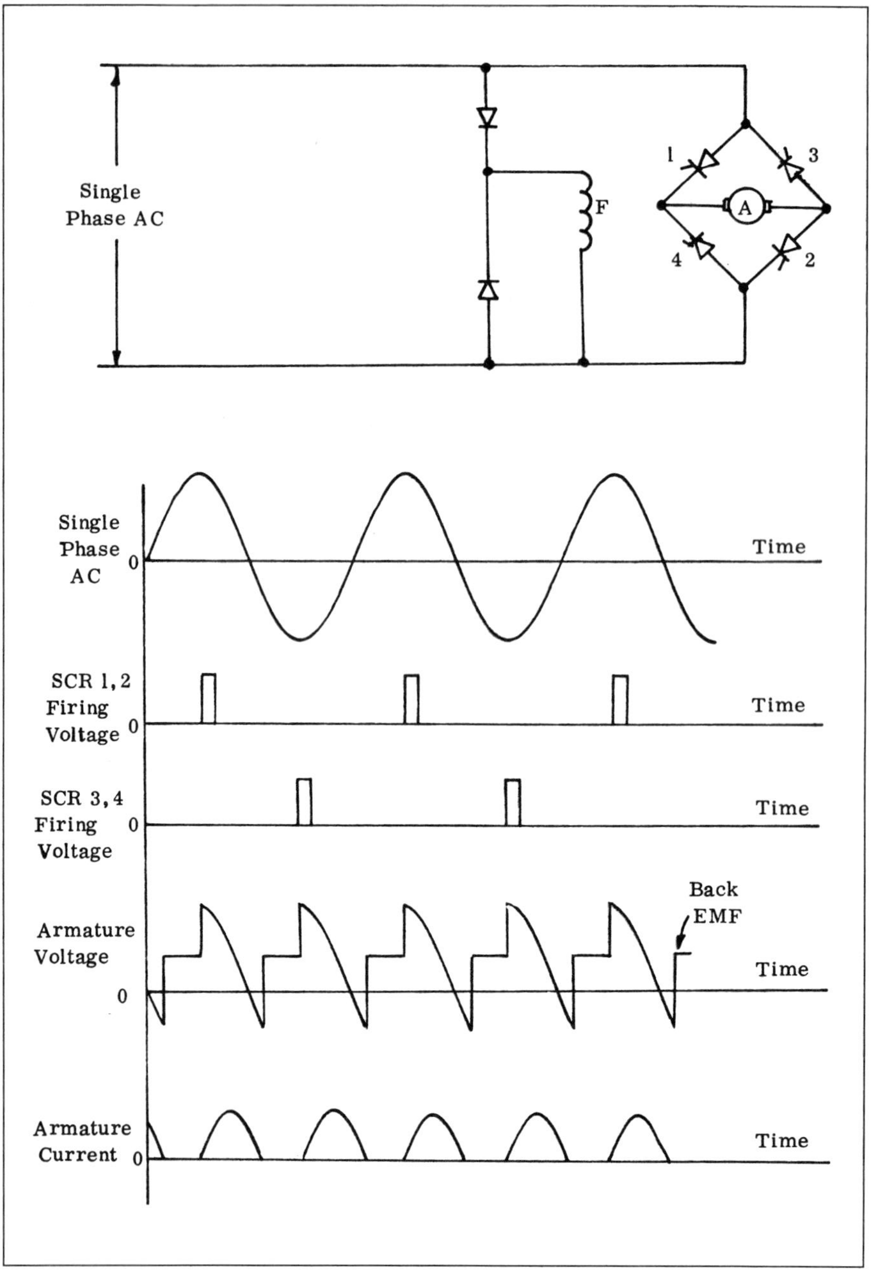

Fig. 6–3 Full-wave motor drive circuit and waveforms

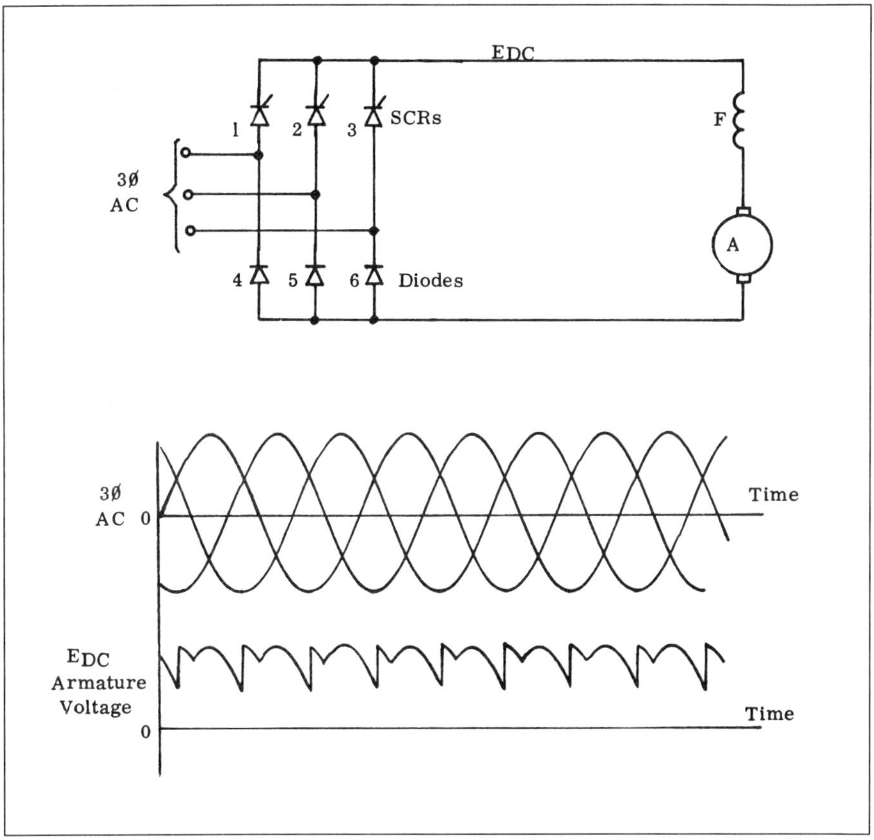

Fig. 6–4 Three-phase, SCR/rectifier, half-controlled bridge and waveforms for motor drive

Pairs of SCRs are fired sequentially in the proper phase sequence to ensure that equal currents flow through each SCR. The SCR gate signals are properly timed and synchronized from the AC line for applications to the SCR gate when its anode is forward-biased. Figure 6–9 shows motor waveforms at various angles of SCR conduction.

Speed and torque control of a DC motor is a function of the voltage applied to its armature and field. In a shunt motor system, the field excitation is supplied from a separate supply and the armature power is controlled from the three-phase SCR bridge. Increasing the armature voltage proportionally increases shunt motor speed up to the full limits of the SCR supply. At a fixed setting of the firing

Fig. 6–5 SCR-driven DC traction motor (Courtesy General Electric Co.)

Fig. 6–6 DC series and shunt-type DC drilling motors (Courtesy Electromotive Division of General Motors Corp.)

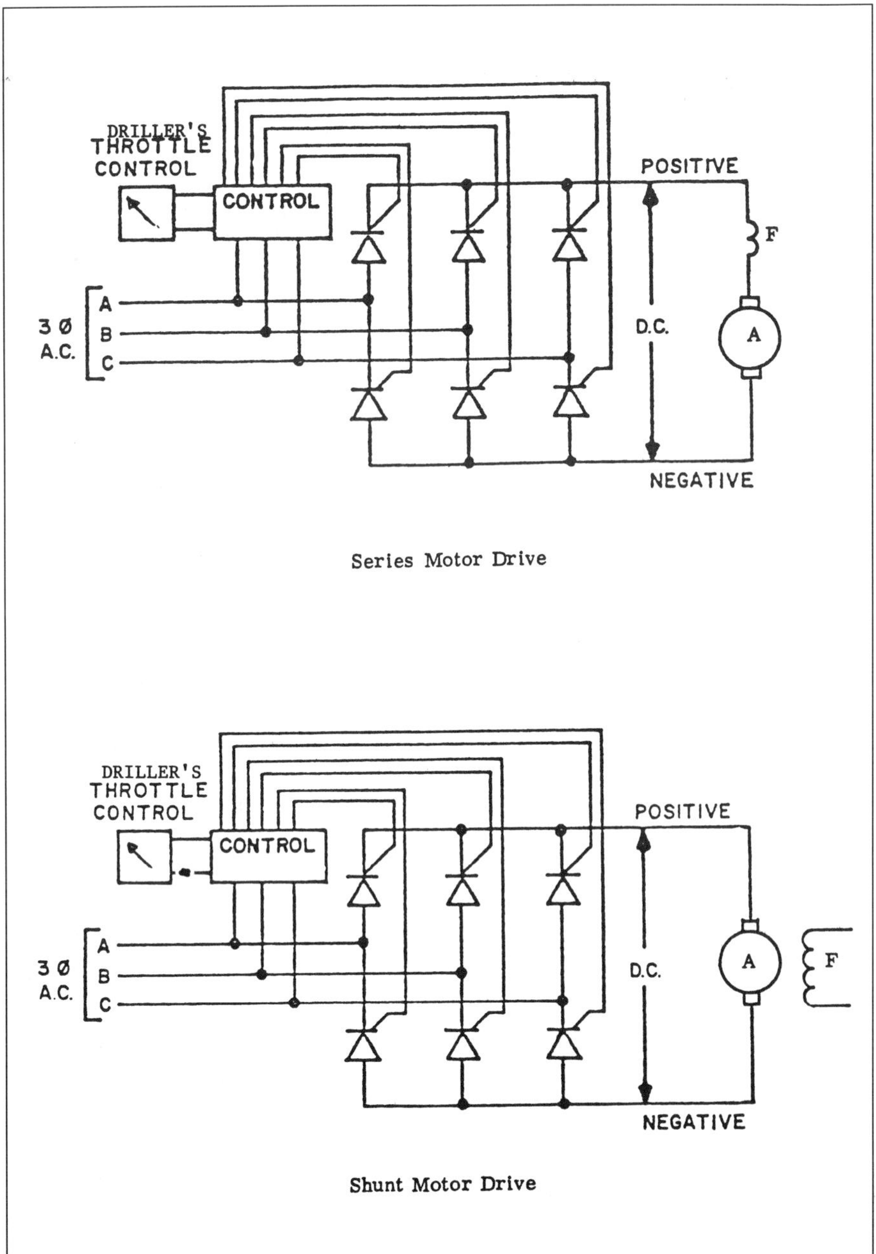

Fig. 6–7 Three-phase full-wave SCR drive schematics

Fig. 6–8 Three-phase SCR bridge assembly (Courtesy Baylor Company)

angle, the shunt motor speed drops about 15% as load is applied to the motor shaft. Adding armature voltage or tachometer feedback into the control system improves speed regulation.

Additional shunt motor speed control is achieved through field-current adjustment over a limited range under constant horsepower conditions. Using SCRs in the field supply allows smooth, easily implemented control.

A series motor does not require a separate field supply, for operation as the SCR bridge supplies the necessary power to meet load demands. Compared to a shunt motor, the series motor speed varies widely with load; careful control and protection are needed to prevent an overspeed condition at light loads. Many SCR manufacturers provide tachometer speed control or use a "speed calcula-

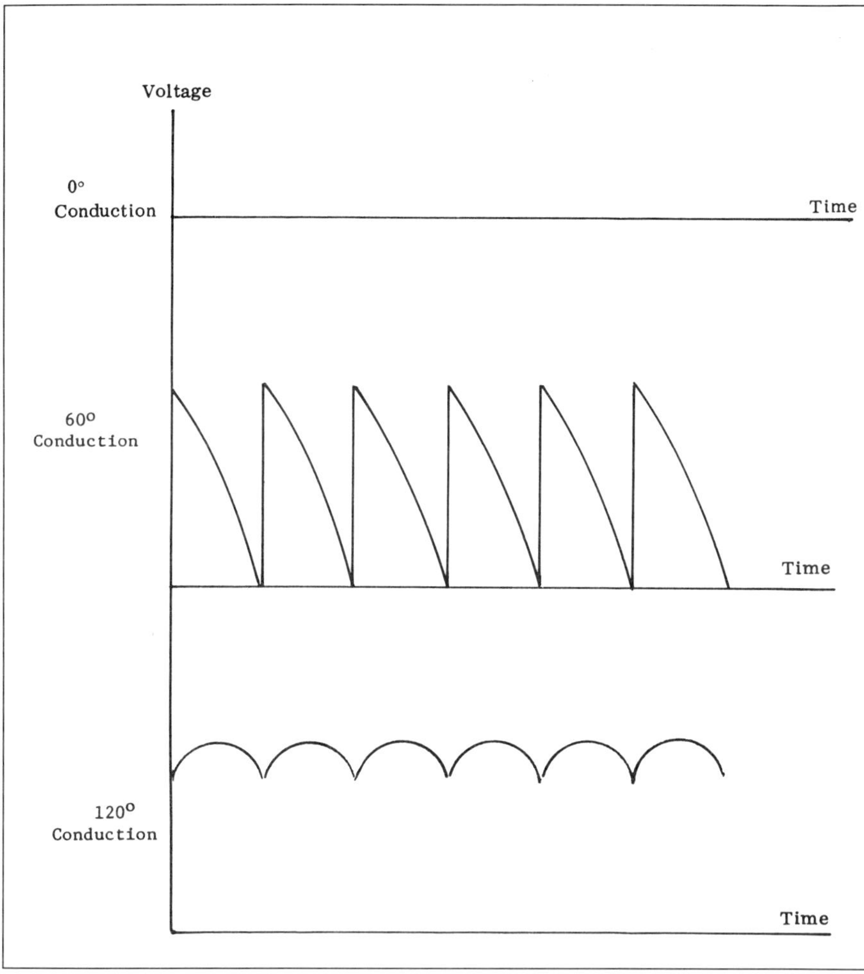

Fig. 6–9 SCR output waveforms under various gate firing angles

tor" in the feedback circuit to compute a signal proportional to speed. In equation form, this equation states:

$$N = K \frac{V_a}{I_f}, \text{ rpm}$$

where K = DC motor constant
V_a = Armature voltage, volts
I_f = Field current, amps (field flux equivalent)

In essence, speed is proportional to the ratio of armature voltage divided by field current. If voltage decreases, current must decrease to maintain a constant motor speed. Conversely, if the voltage increases, the current must increase proportionally.

Dynamic Braking of DC Motors

In many cases, as in drawworks duty on a drilling rig, the motor is actively braked instead of using auxiliary braking or letting the motor coast. One way to achieve this is to convert the motor to a generator and allow its output current to produce a braking torque, dissipating it in a resistor.

Once a shunt motor is up to speed either through SCR-supplied electrical power or mechanical energy supplied from an overhauling load, the motor generates a back-emf or voltage that opposes the applied voltage. If this motor, now acting as a generator, is loaded to reduce its speed, a power resistor can be connected across its armature. The lower the value of the resistor, the higher the load current and resulting braking torque will be.

A series motor is braked similarly, but certainty of residual field excitation or a separate current source for the field must be obtained. In dual-motor systems using series motors, a cross-coupling technique can be used where one armature supplies series field current for the other and vice-versa (Fig. 6–10).

Regenerative Braking

If an SCR drives a shunt motor that is occasionally reversed because of external mechanical loads, the SCRs can brake the motor regeneratively. For example, if a shunt DC motor is at zero speed and the shaft load turns the motor in reverse, the motor acts as a generator and develops a voltage at the armature terminals. When this occurs, the SCRs can be slightly turned on, allowing load current to flow from the motor back into the AC line. This process converts the overhauling mechanical energy into generated electrical energy that is absorbed by the AC line through the SCRs.

Reversing

In many applications, the motor must be operated in both directions of rotation. When reversing is required, it can be accomplished either with contactors or SCRs. Shunt motor reversing can be accomplished by reversing the polarity

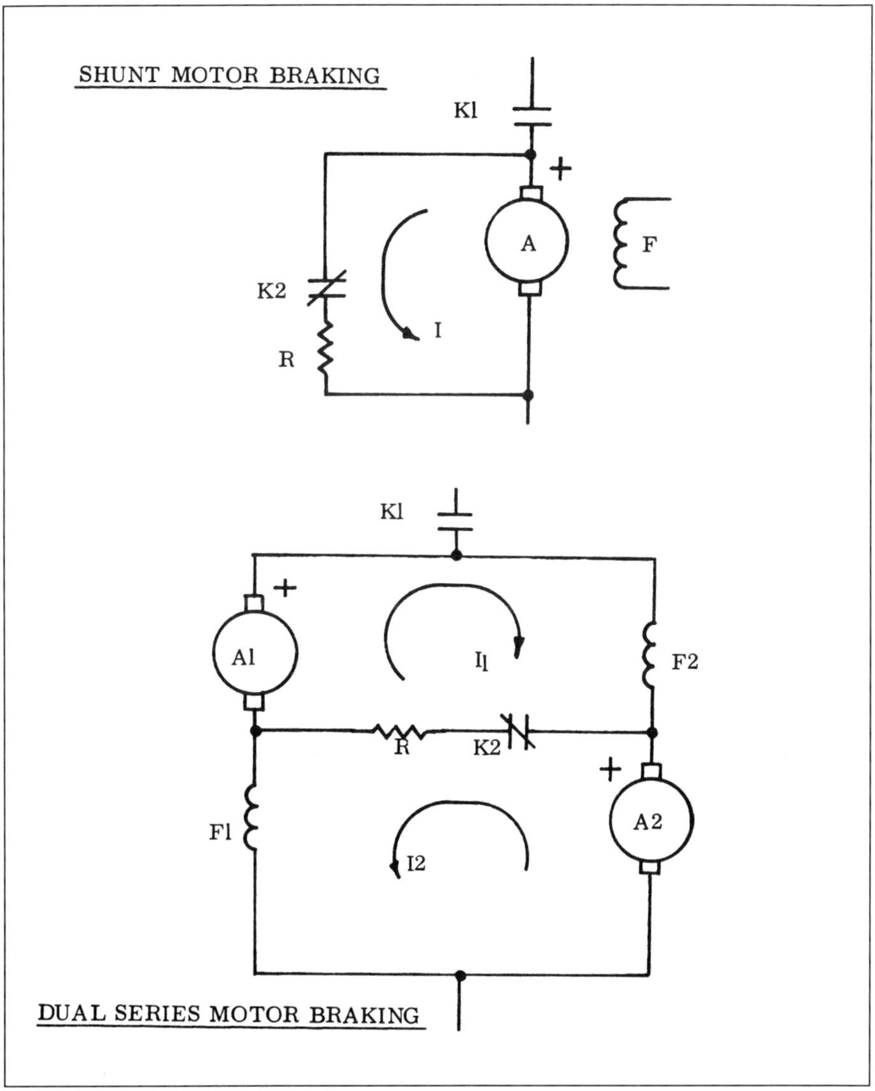

Fig. 6-10 DC motors dynamic braking

of the armature or reversing the polarity of the shunt field.

Figure 6–11 shows several methods to achieve reversing. Armature reversing is achieved faster because of the low electrical time constant of the armature circuit, but it also carries the main motor power. Field reversal can handle relatively low current, but the time constant of the field is high, causing slow response.

Series motor reversing can also be achieved by reversing either the armature or field with respect to the other.

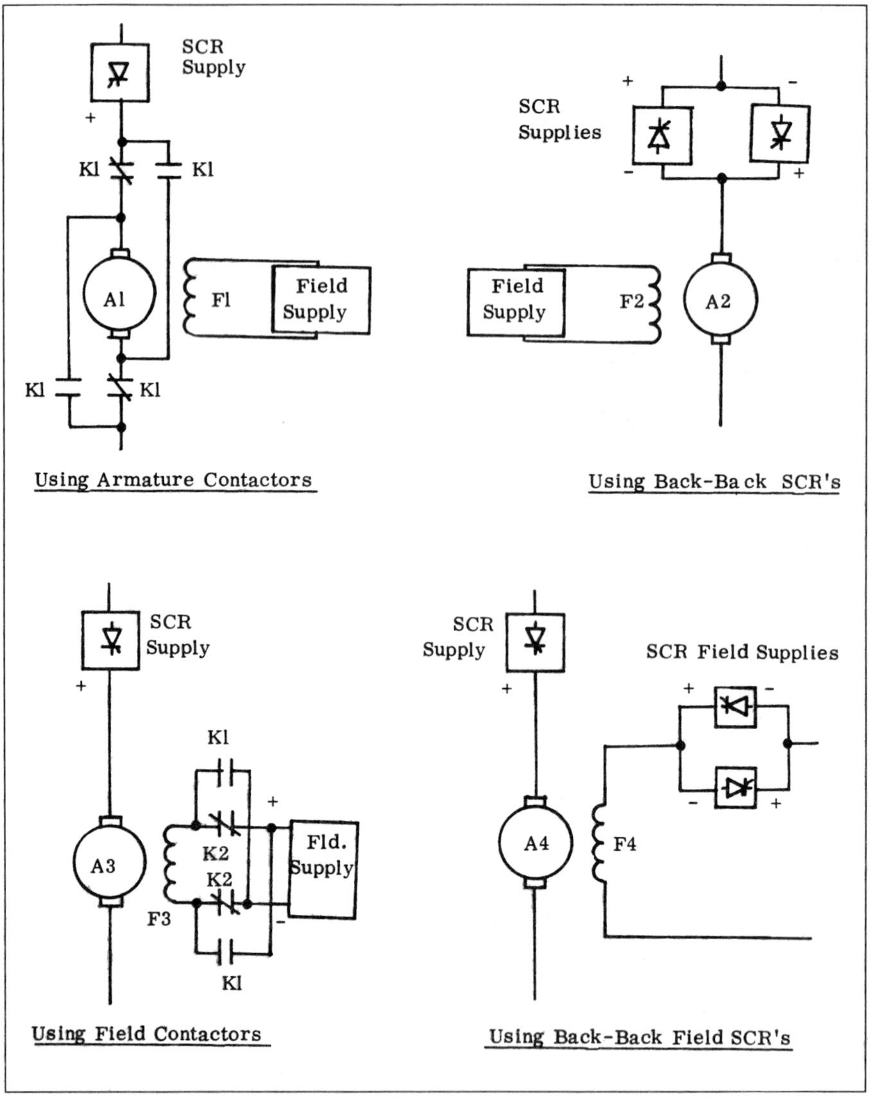

Fig. 6-11 Reversing methods of shunt motors

Load Assignments

One of the biggest advantages of electrical systems is flexibility. Arranging the motor loads where they can be switched from one SCR bay to another allows redundancy should an SCR bay fail. This allows continuous drilling with a minimum amount of downtime. Figure 6–12 shows a typical load assignment schematic with three SCR bays.

SCR bays 1, 2, and 3 obtain their power from the AC bus. Under normal assignment, SCR 1 supplies power to the drawworks motor (DW) through contacts A (B is open), SCR 2 to mud pump motor (MP) through contacts C (D is open), and SCR 3 to rotary table motor (RT) through contacts E (F is open). In the event SCR 3 fails, SCR 2 bay can power motor RT through contacts D (E is open) and

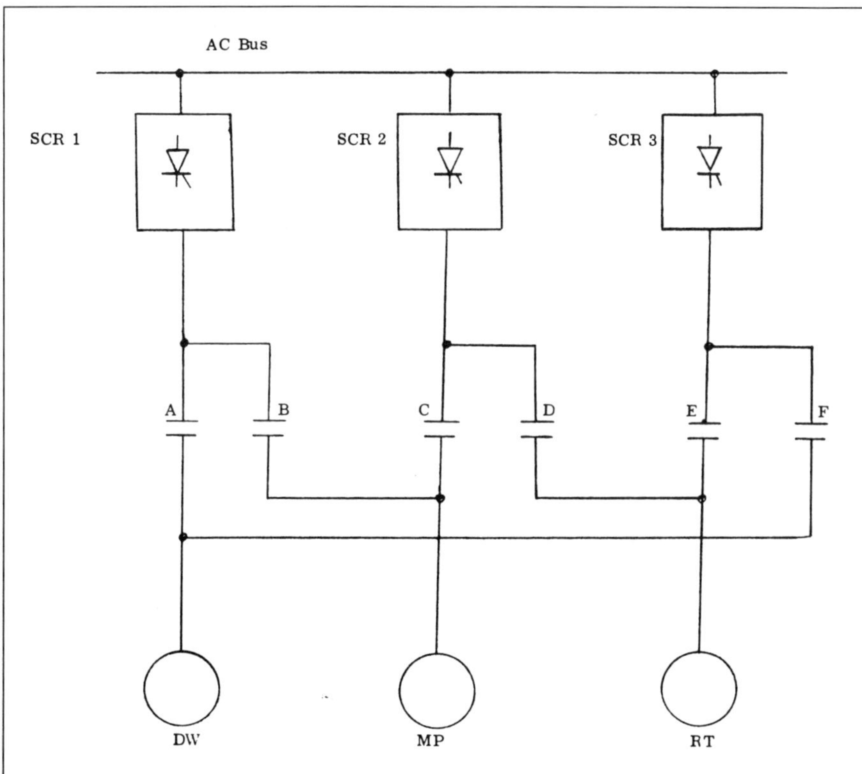

Fig. 6-12 SCR load assignment schematic

SCR 1 can power MP through contacts B. Under these conditions, only two motors could be operated at a given time.

Any SCR bay can fail, and motor reassignment to other bays allows continuous drilling operation while the failed SCR is repaired.

Protective Functions

An SCR system offers additional safety and limiting functions to prevent equipment damage. Series motors, for example, can dangerously overspeed at light loads if their voltage is not reduced. Should this occur, the SCR unit can

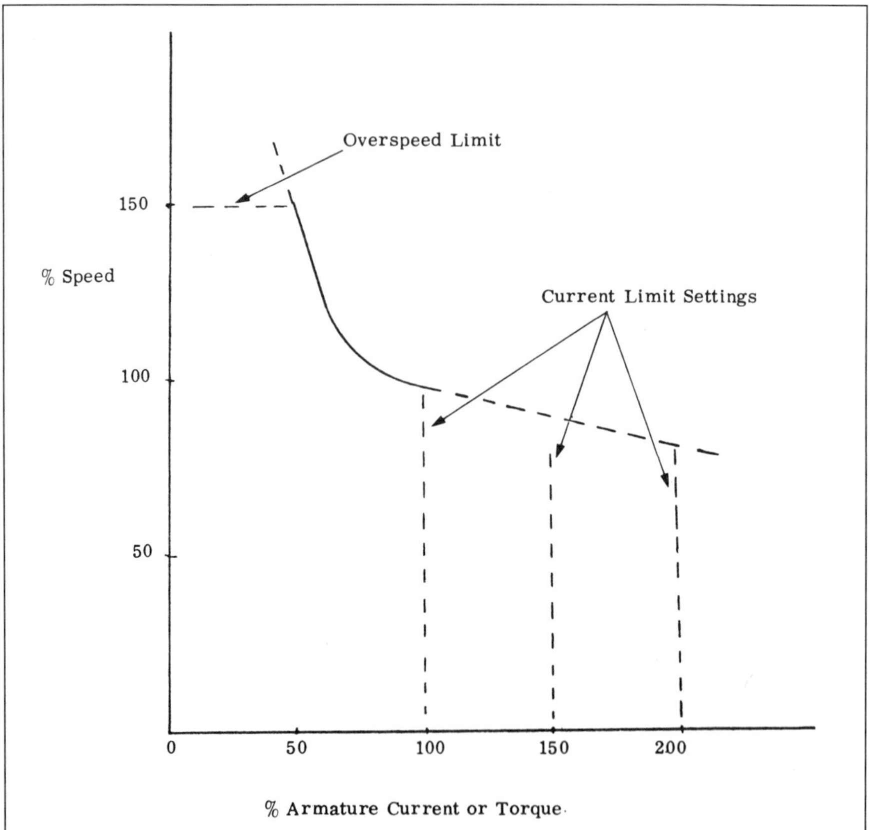

Fig. 6–13 Overspeed and current limit characteristics of a series DC motor operated from an SCR system

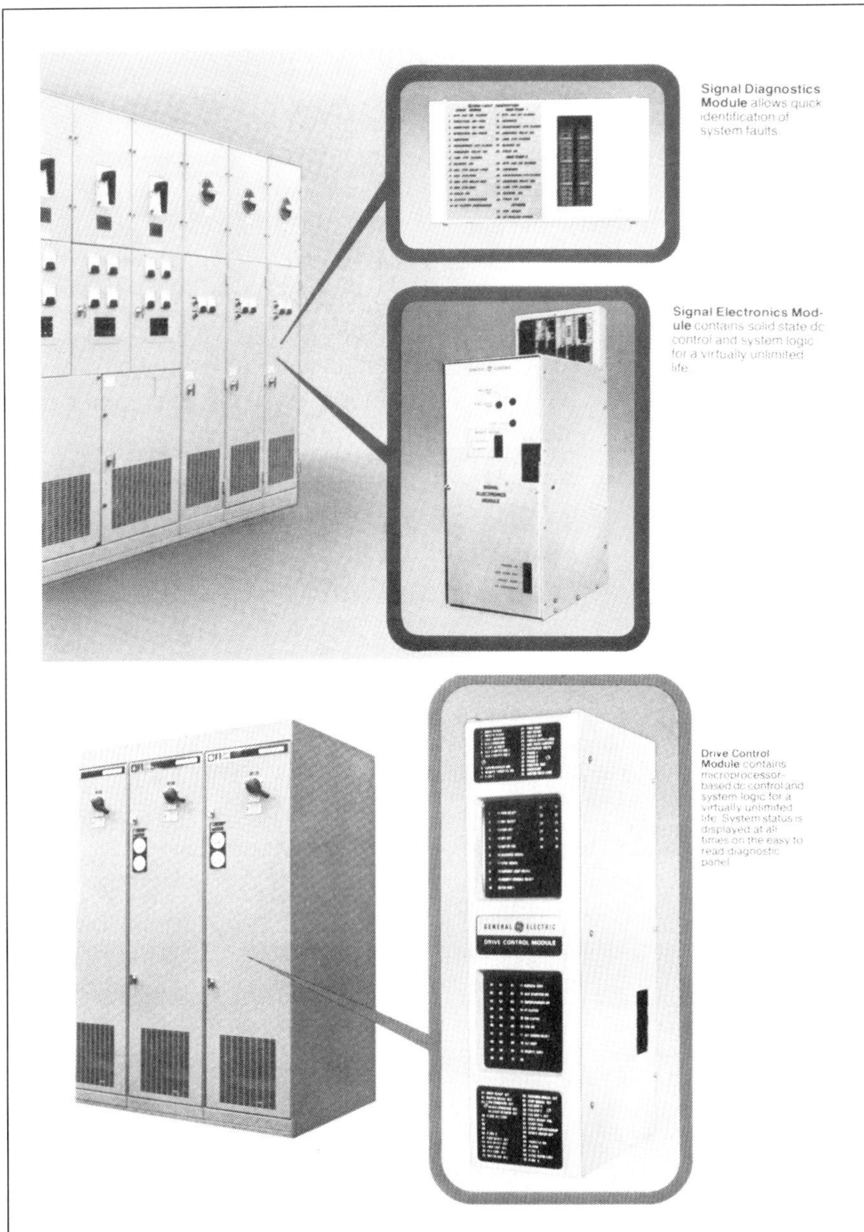

Fig. 6–14 Fault indicating and detecting microprocessor (Courtesy General Electric Company)

detect the overspeed and turn off power to the motor. This is accomplished by sensing the voltage-to-current ratio of the motor, or the speed signal from a tachometer, and turning off the SCR unit.

Overloads resulting from heavy load conditions or acceleration requirements result in excessive DC motor currents. Most SCR units have either fixed or adjustable current limit settings that prevent excessive currents. Current is sensed with a current transducer and, if a threshold level is exceeded, the current limit circuit is activated. Removing the excessive load automatically releases the current limit and allows normal operation to resume. Figure 6–13 shows a motor characteristic with overspeed and current limit functions.

Should a fault occur in an SCR bay logic circuit, a troubleshooting panel or indicator can assist in locating the problem. In some systems, a microprocessor is used to compare fault conditions in the SCR with normal ones and provides alarm and location indication of the fault (Fig. 6–14).

Power Management

Another significant feature incorporated in many SCR systems is power management, or kw or kva limiting, to prevent engine-generator overloads. The circuitry monitors the total number of engine-generator sets paralleled to the AC bus and determines the total capacity of the AC systems at that time. Total power delivered from the AC bus is monitored electronically and compared to a preset capacity level. If the kva exceeds the generator rating or the kw exceeds the engine rating, the power management circuitry limits the output of the SCRs to prevent overload and potential shutdown of the engines.

If an additional engine-generator set is added to the paralleled bus, the net capacity of the power system increases and the power management system adjusts accordingly. Conversely, if an engine-generator set is removed, the total capacity is reduced and the power management circuitry limits the power consumed to this lower level.

On most SCR systems, a "power limit" lamp will indicate that the engine-generator set has reached 90%–95% of its full load rating to the driller. Under these conditions, the power management or power limit circuit will cause the operating SCRs to be limited in their output to keep an overload from occurring. If drilling conditions allow, operating in this mode is ideal from a fuel consumption, maintenance and performance viewpoint.

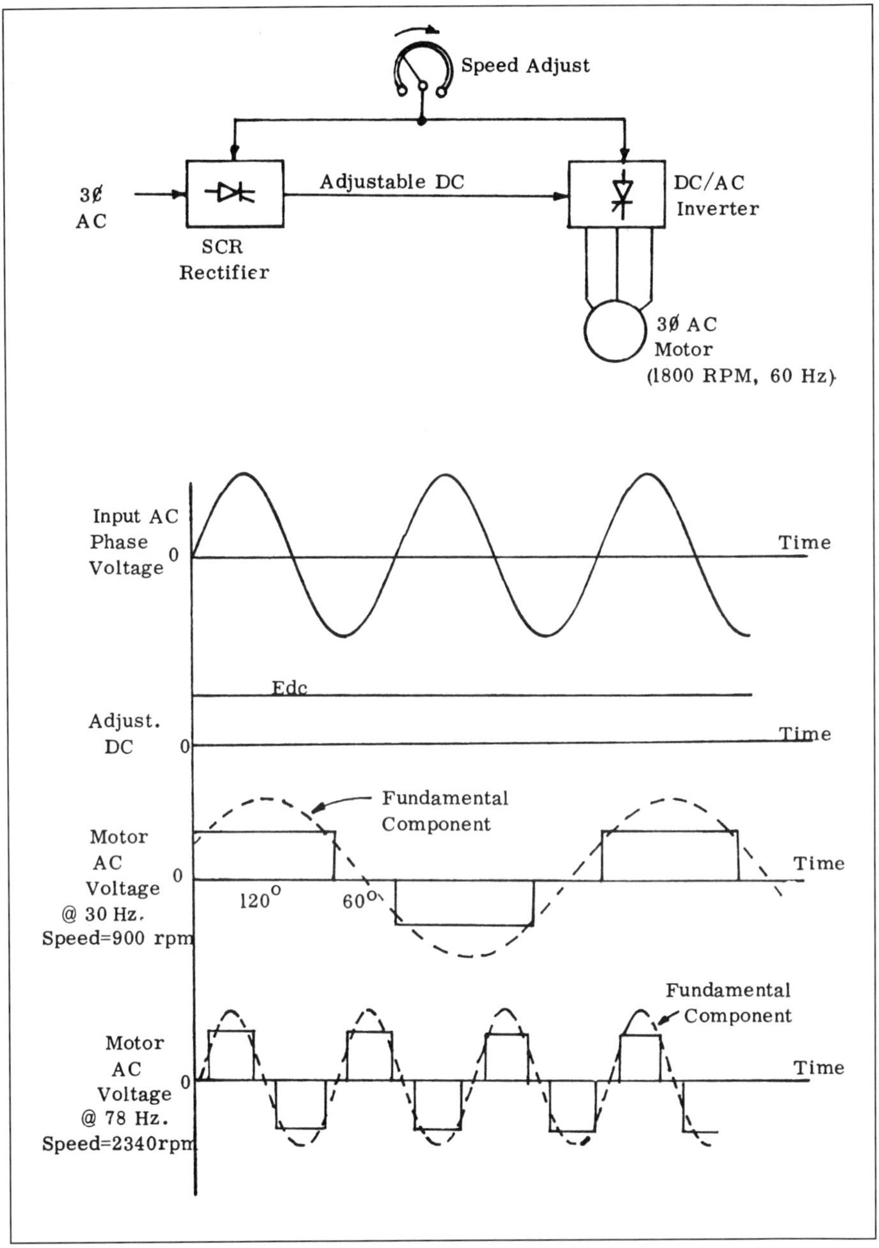

Fig. 6–15 Simple variable frequency motor drive system with waveforms.

SCR AC Drives

The evolution of variable-speed drives has always been toward the system that offers the lowest initial startup, operating, and maintenance costs. The following changes mark the trend of drive systems:

- Engine—DC generator—DC motor (1940s, 50s, 60s)
- Engine—AC generator—SCR rectifier-DC motor (1960s and 70s)
- Engine—AC generator—SCR frequency converter—AC motor (1990s)

As technology reduces the initial cost of a drive system, operating and maintenance costs become the dominant decision factors in selection. An AC motor offers longer life, lighter weight, less maintenance (no brushes or commutator) and lower cost than a DC motor. However, the technology to produce an SCR inverter at low cost competitive with an SCR rectifier has not been accomplished to date. These technical and economic hurdles will be overcome so that the drilling industry can benefit from the AC drive advantages just as the chemical and process industries have.

AC drives offer several advantages. The block diagram of Figure 6–15 shows a simple variable-frequency converter operation. Speed control is set by the adjustment potentiometer that feeds a control signal to both the AC/DC rectifier and the DC/AC inverter. Several characteristics of both the control and motor must be analyzed in order to understand its operation.

The synchronous speed of an AC motor is directly proportional to its frequency:

$$N = \frac{(120)(f)}{P}, \text{ rpm}$$

where N = speed in rpm
f = frequency in Hz
P = number of motor poles

For AC induction motors, its actual shaft speed is slightly less than synchronous speed because of its inherent slip characteristics. However, the following principles of variable speed apply.

The motor voltage must be maintained in direct proportion to its frequency. A motor is a magnetic device similar to a transformer; it has a certain maximum volt-second energy product inherent in its design. If this is exceeded, the motor primary will magnetically saturate and excessive currents will flow in the windings without producing any additional torque. Therefore, to produce comparable torque without overheating the motor over its speed range, a variable frequency control must maintain approximately constant volts-per-cycle to operate the motor properly. For example, a standard motor designed to operate at 460 VAC, 60 Hz, and 1750 rpm requires only 46 VAC, 6 Hz, to produce 175 rpm and still produce full-rated torque.

The best motor voltage waveform is sinusoidal. Any distortion of a waveform from a sine-wave represents harmonic voltages which do not produce any useful torque but contribute to motor heating. However, a motor winding and its rotor inertia represent a frequency filter to higher frequency harmonics, thereby allowing some waveform distortion without affecting motor performance. In most practical frequency converters, the applied voltage is not sinusoidal but an approximation of the sine wave fundamental. The result is some torque loss which is small compared to the torque produced by the fundamental frequency.

The voltage waveforms in Figure 6–15 demonstrate the developed voltages in an inverter. Other waveforms produced by inverters are called "quasi-sine waves."These waveforms do not possess "even" harmonics (2,4,6..multiples) but possess "odd" harmonics. In a three-phase system, no third harmonics or any multiples (9th, 27th,etc.) exist due to the 120° voltage displacement. Other odd harmonics (5th, 7th, 11th, etc.) may exist, depending on the type of waveform. The higher the harmonic frequencies, the less effect the waveform has on the motor torque.

An AC synchronous motor always runs at its synchronous speed without any slip under load. An AC induction motor runs at some slip, or less than synchronous speed, depending upon the motor load and rotor resistance.

An AC motor drive offers improved motor reliability and provides significant application improvements for the drilling contractor.

1. Motor reversing of the drawworks, rotary table and top drive can be achieved at the control level and not at the power level by reversing the phase sequence of the output voltage to the motor.

2. AC induction motors can be used on the drill floor with more confidence since no brushes or commutator can produce sparks in the hazardous areas. Explosion-proof enclosures can also be obtained with these motors.
3. The brushless AC induction motor is less susceptible to damage by a corrosive atmosphere or environment typical of oil drilling since it is usually housed in a totally enclosed, fan-cooled, or explosion-proof (XP) enclosure.
4. Speed regulation is inherent in the motor selected and the frequency stability, not in the control system.
5. Regenerative control and braking due to an overhauling drawworks is automatically controlled with an AC drive system.

7 DRILLING CONTROLS

Once an electric drive system is put into service, the driller's primary control is the drilling console. Typically, this console is a rectangular steel enclosure with instruments, switches, and speed control knobs (Figs. 7–1, 7–2).

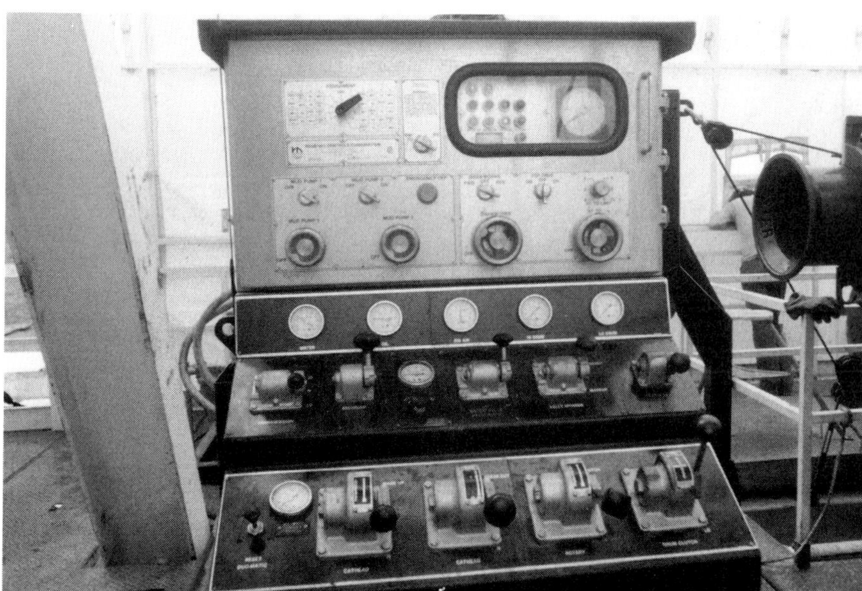

Fig. 7–1 Typical land drilling console (Courtesy Ross Hill Controls Corp.)

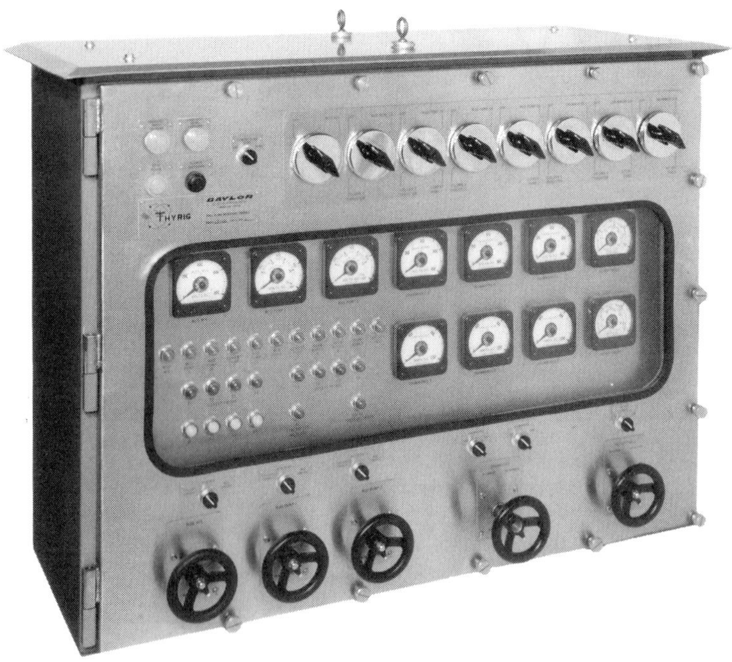

Fig. 7–2 Typical offshore driller's console (Courtesy Baylor Company)

The driller's console typically provides these functions:

- •All control cables interconnecting the console with the SCR/generator power system;
- •Speed control adjustments of the mud pumps, drawworks, rotary table, top drive, and cement pumps;
- •Selective reversing of the drawworks, rotary table, and top drive;
- •Dynamic braking actuation of the drawworks;
- •Switching assignments of the motor loads to a specific SCR control bay;
- •Emergency stop pushbutton;

- •Indicating lamps showing alarms and system status of generators and other equipment;
- •Ammeters and voltmeters of drilling motors' currents and voltages;
- •Rotary table torque in percent of rating;
- •Engine-generator power capacity being used;
- •Selection of drawworks hand throttle or foot throttle.

After the generators are turned on and manually switched to the system AC bus at the generator control bays, the SCR bays' circuit breakers are closed. This allows the system to be controlled from the driller's console. All status lights on the driller's console are checked to determine proper operation. Assignment selection of the motor loads to the appropriate SCR bay is done by the driller from the console, and forward or reverse operation of the drawworks and rotary table (top drive) is selected. Then, the mud pump's speed is increased to the desired level by manual adjustment of the throttles, and the speed of the drawworks is adjusted to the desired level.

When an independent rotary table or top drive is used, its speed is adjusted to the desired level. Visual observation of the meter showing power demand required on the system should not exceed the generated kva or kw.

Another control the driller uses is the foot throttle, which controls the drawworks speed when the hand throttle is not being used. Transferring speed control to the foot throttle is achieved with a selector switch on the driller's console or by cracking the hand throttle slightly and depressing the foot throttle. This allows the driller to perform additional control functions while maintaining drawworks speed control with his foot.

The foot throttle (Fig. 7–3) is a signal transducer housed in a steel enclosure to protect it from hazardous environments found on the drill floor. As the foot throttle is depressed, the signal to the drawworks SCR bay calls for increased drawworks speed.

Some systems require a separate control console for cement pump control. This control is coordinated with the main console to allow one of the SCR bays to operate drilling functions or the cement pump. The cement pump control console is typically located at the cement pump site and is similar to Figure 7–4.

A mast-raise console is sometimes used for the rig- up/rig-down process instead of the main control console. Control transfer is made from the driller's main console to the mast-raise console, where the drawworks provides power for the mast-raising event.

Local mud pump control is sometimes required (Fig. 7–3). It consists of motor speed throttles and selector switches for either local or remote operation.

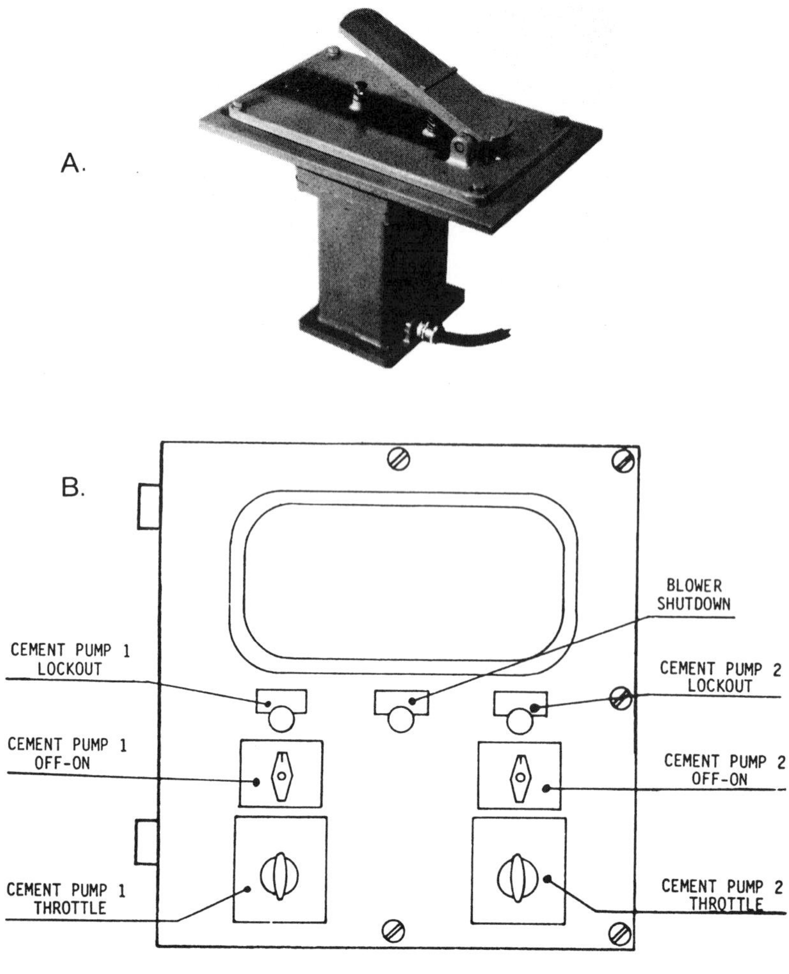

Fig. 7–3 Foot throttle and cement pump console (Courtesy of Baylor Co.)

Fig. 7–4 Mud pump operator's console (Courtesy Baylor Co.)

Electric Brake Controls

Eddy current brakes used on drawworks obtain their control power from a variable DC power source using either SCRs or power transistors. This DC control current passes through the brake coils to produce braking torque proportional to this current. Thus, by varying the DC current from zero to maximum, the brake produces torque up to its full rating.

The SCR brake control receives its source of power from the three-phase, 600-VAC system bus and, through transformers, reduces the voltage to 220 VAC, three phase. This AC is delivered to the SCR brake controller, consisting of a six-device SCR bridge and control module. Its control signal is received from a hand throttle located on the drill floor adjacent to the driller's console (Fig. 7–5).

The throttle is a variable-voltage transformer whose output is an AC voltage proportional to the position of the throttle handle. This AC voltage is conducted to the control module within the control enclosure where it is properly conditioned to generate an SCR gate signal. This gate signal advances its firing angle proportional to the throttle signal to produce a DC voltage from the SCRs to the brake coils. Reducing the position of the throttle calls for less brake coil current.

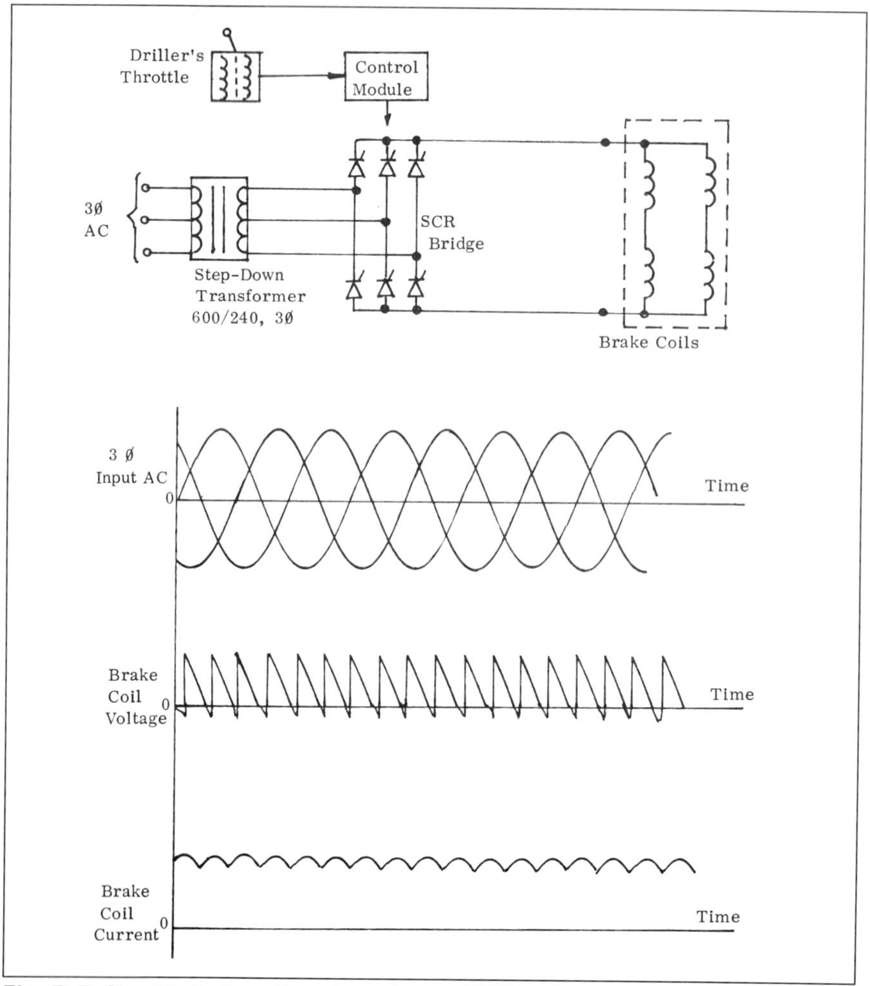

Fig. 7–5 Simplified schematic and waveforms of SCR eddy current brake control

The coil energy "regenerating" back into the SCR supply through the SCRs results in a fast decay of the braking torque demanded by the driller (Fig. 7–6).

A further innovation in brake controls is a system using pulse width modulation (PWM) to control brake power. This improvement, using transistors, provides several features over the SCR system. First, throttle control position is linearly proportional to DC brake voltage, thereby giving the driller better control of brake torque. Second, controls are housed in a convection-cooled enclosure for easier maintenance, longer operating life, and lighter weight. Finally, battery backup power is provided in the event of AC power failure. If AC loss should occur, the batteries would provide control power to permit continuous brake power control within the capacity of the batteries.

The PWM control rectifies the incoming three-phase AC to a fixed DC level of approximately 250 volts DC. This DC voltage is then switched through the power transistor at a variable-duty cycle to control the average DC voltage to the brake coil. At low-duty cycles, the average voltage is low and increases as the "on" duty cycle increases. At 100% duty cycle, the transistor conducts constantly and the full DC rectified voltage is applied to the brake coil. Figure 7–7 shows two duty cycle waveform examples of 50% and 90% with a simplified schematic of the control.

Starting Large AC Motors

An occasional need may exist for controlled starting of large AC motors on the SCR system AC bus. Since the starting currents of AC motors can be five to eight times their full load rating, some means of limiting this inrush is needed to protect the system AC bus generators. Typical controls of this type are called reduced voltage starters (RVS).

Some RVS's use a resistor-contactor method of starting motors. The resistors are fully in series with the motor when first started and, with the aid of timers, are sequentially shorted by the contactors until the motor is at full speed. While this effectively limits the current, the power resistors and contactors are large, bulky, and costly.

A more common technique is motor starting with autotransformers. When the motor is first started, the motor is connected to the low-voltage transformer taps. As the motor comes up to speed, the motor is transferred to higher transformer taps with contactors or switches until full voltage is achieved. These RVS's are relatively heavy and expensive (Fig. 7–8).

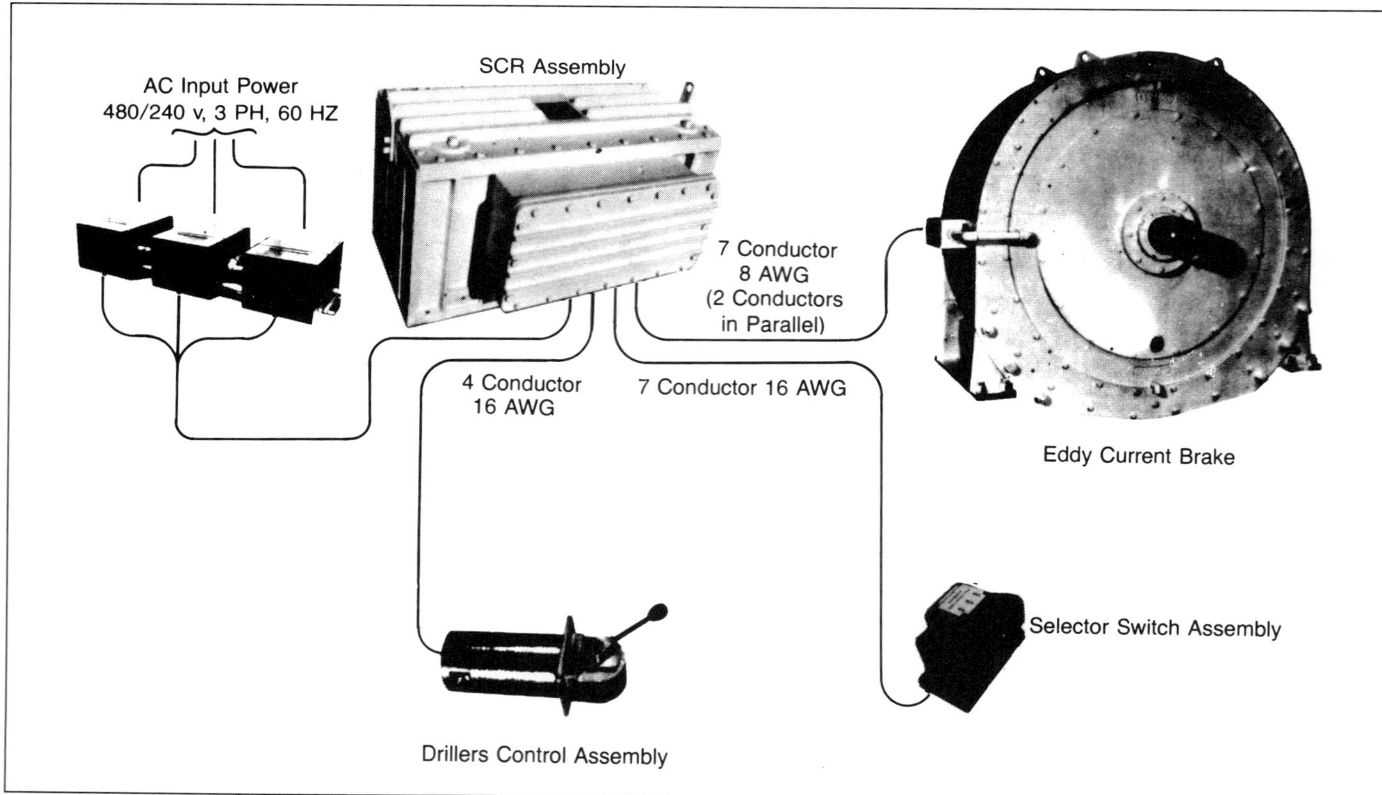

Fig. 7–6 SCR brake control system simplified diagram (Courtesy of Baylor Co.)

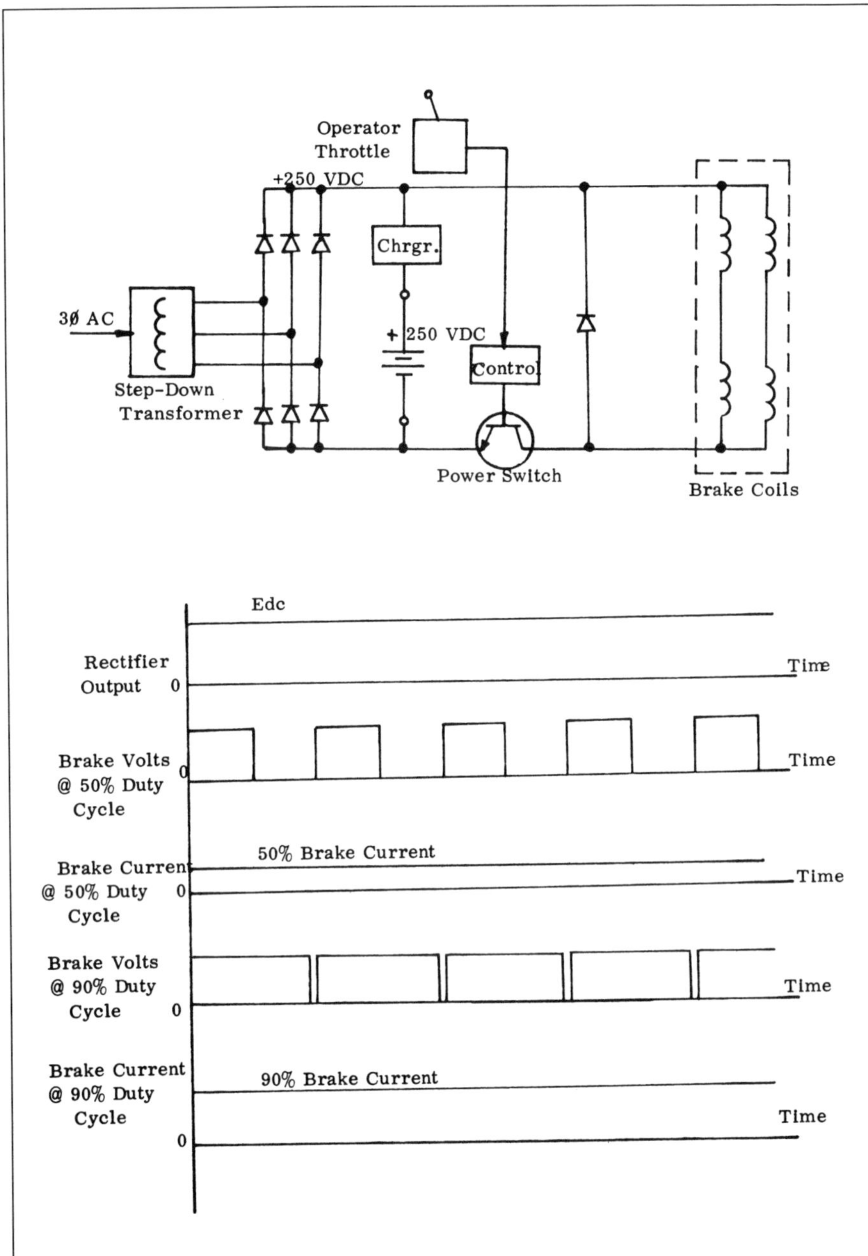

Fig. 7–7 Pulse width modulated (PWM) brake control (Courtesy Baylor Co.)

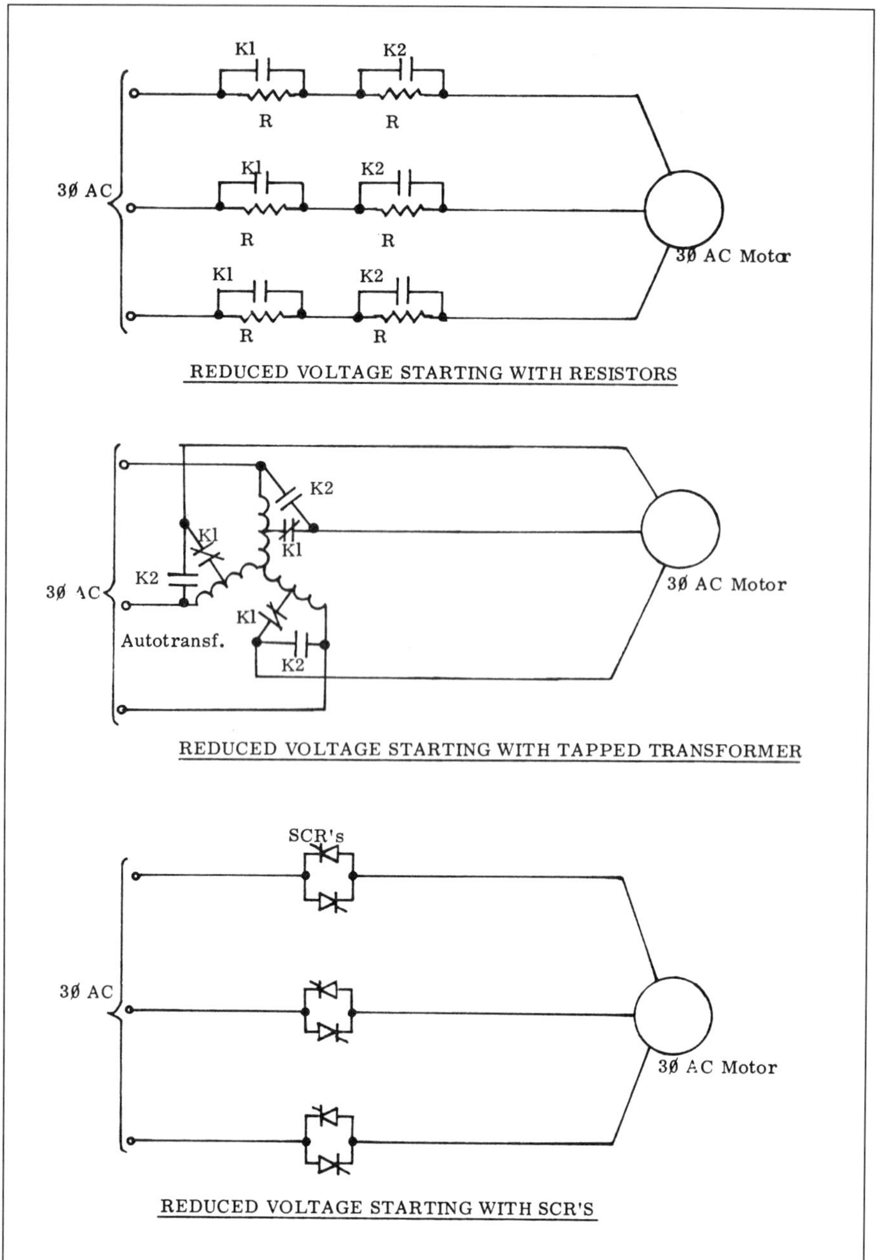

Fig. 7–8 Reduced voltage motor starters

The most modern RVS is an AC-SCR starter in series with the motor. Current is limited by retarding the conduction angle of the SCR and slowly advancing it until full speed is achieved. This method offers smooth starting in a small package but is slightly more expensive than other techniques (Fig. 7–9).

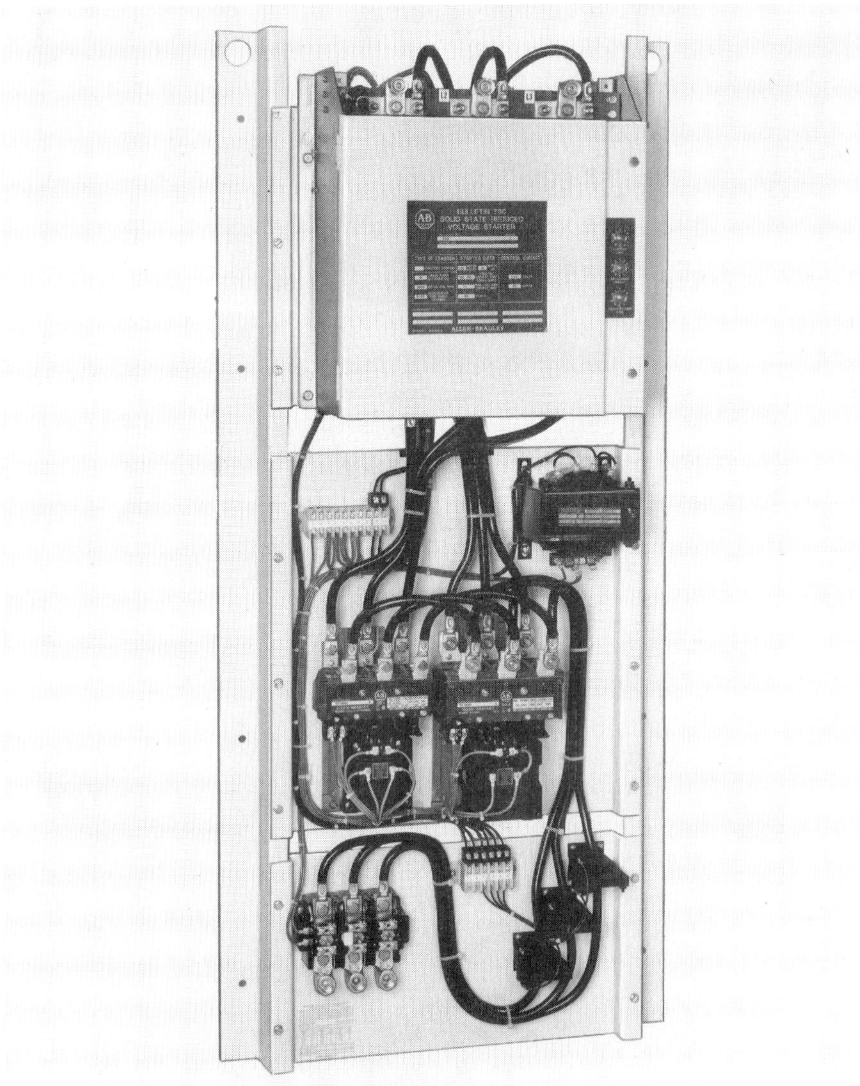

Fig. 7–9 Solid state reduced voltage motor starter (Courtesy Allen Bradley Co.)

Thruster and Ballast Controls

Thruster speed and propulsion control on offshore drilling vessels is achieved primarily by one of two methods: Either the drive motor to the thruster is a variable-speed motor operating from an SCR DC or AC control, or the drive motor is a constant-speed AC motor with variable-pitch propellers.

Thrusters serve several functions. Large thrusters provide the main propulsion power for transporting the drilling vessel from one location to another; small units are used to maneuver the vessel in a dynamic positioning or docking operation.

Horsepower sizes of these thrusters vary from several hundred to several thousand and are usually geared or chain-driven from the electric motor to operate at top speeds from 150–400 rpm. In many cases, two- or three-drive motors are mechanically coupled to the main propulsion thruster to achieve the desired horsepower and redundancy.

SCR DC controls are the most common type of variable-speed drive, with both armature and shunt field control of the motor to achieve the desired torque, speed, and load sharing (Fig. 7–10). AC drives will provide other application advantages as they are introduced in future thruster systems.

Once the floating vessel is positioned at the drillsite, it should be kept level as various deck loads are shifted from one location to the other. Controls such as those shown in Figure 7–11 adjust the ballast to maintain a level attitude.

Jacking Controls

One of the functions on jack-up rigs is raising or lowering the drill platform above the water surface. Typical systems use high-slip AC induction motors operating at fixed speed from a start-stop control panel. Others use hydraulic motors in a similar fashion. Several schemes perform this function, but one popular technique is a variable-speed drive.

A typical jacking system operates at top speed of approximately 1 ft/min as the platform moves vertically along the support columns or legs. Typically, 250–500 HP per column will be required to achieve the load and speed requirements of jacking.

Electric SCR drives operating through appropriate speed reducers are ideally suited for this application. Compared to hydraulic-drive systems with their associated piping and maintenance problems, SCR drives provide easy installation, application flexibility, and easy control.

Fig. 7–10 Thruster control console (Courtesy Baylor Co.)

Fig. 7–11 Ballast control console (Courtesy Baylor Co.)

For example, four separate 100-HP DC motors can be used per column and operated from a single SCR control. The output shafts can be mechanically coupled to the jacking column and operated at the same speed. The series motors share load and contribute their proportional share of torque to powering the system.

Radio Frequency Interference

As SCR power controls become more prevalent in drill rig applications, the radio frequency (RF) emissions they generate become of greater concern to rig designers. This RF radiation can interfere with the proper function of the other electronic equipment on the rig.

On most land drilling rigs, the SCR and generator control equipment are the only electronic gear in the area. At times, the land rig system has two-way radios for voice communications between the tool pusher and other company personnel, or radio receivers for news and entertainment. Interference of this communication equipment can occur if the SCR equipment is near, not enclosed in metal cabinets, or if the radios have no filtering against line noise. In general, land rig units do not have enough critical electronic equipment to create much concern for RF interference.

Offshore drilling systems such as drillships, semisubmersibles, platforms, and jack-ups have more sophisticated electronic equipment than land rigs. Included in the equipment are thruster controls, environmental instrumentation, computers, navigational equipment, and satellite communications. Equipment location, installation wiring, and filtering must be considered to ensure proper operation of all electronic systems. The following information provides the rig designer and electrician more background into the causes and effects of electromagnetic interference.

Definition

Electrical and electronic equipment and systems are affected by electromagnetic energy. When a malfunction results because of this energy, the equipment or system is susceptible to electromagnetic interference (EMI).

EMI is classified by its mode of travel—radiated or conducted. Radiated interference is unwanted electromagnetic energy that emanates from any equipment or interconnecting wiring. Conducted interference travels through equipment, cable, or interconnecting wiring. Conducted interference travels through equipment by external connections.

EMI also is classified by its frequency characteristic—narrowband or broadband. Narrowband interference is unwanted electromagnetic energy which is present only at one frequency or over a narrow frequency range. Radio-type devices operating at discrete frequencies are common sources of narrowband interference.

Power Line Sources of EMI

Power lines are potential sources of radiated EMI; in addition, equipment and systems that are metallically connected to these power lines are sources of conducted and radiated EMI. Power-line noise, practically random in behavior, is of the sum of low- and high-level sources.

For example, radiation noise on conventional transmission lines is enhanced by the low-level sources normally present in the form of line impedance discontinuities (branch points, distribution transformers, switch gear, and power-factor correction components) and by abnormal factors (poor or corroded connections, defective insulators, and contact with noncurrent-carrying metal parts). High-level impulses, which add to these low-level sources, include random pulses attributable to switching operations, faults, surges, lightning discharges near the power line, and periodic impulses due to rotating machinery and rectifier circuits connected to the line.

Rotating Machinery Sources

As mentioned, rotating machines are major sources of impulse noise conducted and radiated by power lines. The direct radiation effects, however, are confined to the immediate vicinity of the machine.

Direct-current machines contribute brush noise from surface contact transients and sparking as well as commutation transients. AC machines are also subject to brush/slip-ring noise and harmonic generation due to electrical and mechanical asymmetry. Noise components fall off at about a 24 decibel/octave slope at frequencies above a few hundred kHz.

SCR and Rectifier Sources

SCR (thyristor) and rectifier equipment produce EMI primarily because of the fast-switching current changes in their circuits. When rectifiers turn off or SCRs turn on or off, the event is usually accompanied by a sudden change in current. This rapid change creates high-frequency harmonic current components which produce EMI, either conducted or radiated from the equipment.

Other sources of EMI are ignition systems, TV and radio sources, fluorescent lights, and intentionally generated sources such as radio transmitters and radar.

Defining Susceptibility

EMI can be diagnosed by noting its effect on components, either by viewing the component as isolated or as part of the system. If the system depends on each of its components, the system criterion determines the allowable susceptibility of the system.

Electronic systems contain components and equipment joined by cable paths that carry electronic signals as they are processed. Cables that carry the signals are not considered "critically susceptible" since their function is not affected by EMI. However, active electronic components may be devices that contain sensitive input signal paths. These sensitive inputs may become "critically susceptible" to unwanted signals which will cause them to malfunction. A large unwanted signal will be readily detected and may even damage the exposed device.

Besides receiving weak signals, electronic components perform a variety of functions which may or may not be used simultaneously, such as voice transmission, data transmission, command and control, navigation and ranging, radar, television, and telemetry. Some systems use one or many of these functions, with satisfactory performance becoming more difficult as the number of functions increases.

Certain methods can establish acceptance ratios that can apply to systems with different parameters from those assumed. In other cases, either published values or rather arbitrary choices have been used. Susceptibility analyses facilitate interference prediction.

Prediction by systematic analysis, however, is not always possible; experimental methods may have to be used. The most useful test procedure simulates actual use most accurately. The practical approach is the test which essentially guarantees success in the field under the most stringent conditions. The intensity of the electromagnetic fields and the kind of waves that can be encountered are known. Tests using the upper limit of expected field strength and an extreme type of waveform provide the required guarantee.

Standards for evaluating system susceptibility are generally unavailable. An accurate evaluation of system susceptibility requires much judgment in anticipating the system's operating environment. A proper evaluation of system susceptibility considers the system performance specifications, then describes the environment.

Intrasystem susceptibility refers to possible performance degradation of various components of a system due to incompatibility of various components in that system. Data are obtained from measurements of radiated and conducted noise from the various subsystems. These are compared with the susceptibility data for the various subsystems, with a knowledge of the subsystems' layout in the final system installation. Performance degradation, because of intrasystem incompatibility, then can be predicted.

In identifying the overall system's susceptibility, information must enable one to estimate the effectiveness of the system in given operational situations. The problem is interpreting the degradation of system performance—as given in terms of technical parameters—to provide information on the system's effectiveness under operational conditions.

Performance criteria on overall system operation presumably will become more explicit and quantitative in the near future. At present, the goal should be to report the effects of environmental conditions on performance as objectively as possible.

Solutions to Attenuating EMI

Related groups of equipment or subsystems may be made electromagnetically compatible by any combination of three fundamental approaches: The interfering signal level may be reduced to a tolerable value; the receiver susceptibility may be reduced; or the attenuation of the path or paths over which interference is transmitted from source to receiver may be increased.

In turn, the attenuation of the propagation path may be increased by either of two basic approaches, or a combination of them. If the path is conductive, filters may be inserted in the lines between source and receiver. Interference signals generated by electromagnetic fields may be attenuated more effectively by electromagnetic shielding, either of the source or the receiver.

A source of radio frequency (RF) energy is enclosed with a shield to contain most of that energy. A properly designed shield permits transmission of only a small amount of energy from the inside to the area outside the shield. Conversely, a shield enclosing a receiver protects the receiver against radiation from many sources and prevents the transmission from outside the shielded region to within it.

Radio frequency interference (RFI) filtering differs considerably from ordinary filtering. Standard, well-established filter theory is based on idealized conditions such as matched impedances and perfect components. RFI filters rarely can be designed with such simplified assumptions.

Bonding is a fixed union of two electrical conductors which has a fixed resistance to a current passing through it. From an EMI viewpoint, in good bonding, the resistance is negligible for the application considered. Especially important in the reduction of RFI are the connections inground system, other common system connections, and those that join electromagnetic shields together.

The literature on the subject is scant—electrical bonding has appeared so simple and commonplace that it has received little serious analysis or explanation. Although metal bonding has been studied thoroughly in its physical aspects (welding, brazing, etc.), good electrical bonding has not been given enough attention. The relatively high levels of electrical energy, plus the necessary structural strength required, almost ensure good bonding.

Installation Suggestions for Generating and SCR Equipment to Minimize Radio Frequency Interference

1. Install AC generators, generator controls, and SCRs as closely together as possible, preferably in the same room.
2. Rooms containing generators and SCR controls should be steel constructed with doors metallically connected to door frame. This reduces radiated EMI.
3. Cables from each three-phase AC generator should be run together in individual steel conduits and slightly twisted in a spiral, if possible. Some cancellation of radiated EMI also results from this technique.
4. Cables from each SCR bay to each DC motor should be run together in individual steel conduits and slightly twisted in a spiral, if possible. Some cancellation of radiated EMI also results from this technique.

5. All power transformers connected to the AC generators and SCR equipment should be enclosed in steel cubicles along with copper wire mesh screen at all ventilating openings. This reduces radiated EMI.
6. All hotel and lighting distribution panels should be located outside the SCR/generator room. The transformer supplying this voltage from the generator should be located in the SCR room and its secondary should have an EMI filter for conducted EMI. Cable runs after the filtering should be through a tight-fitting bulkhead to the hotel and lighting distribution center.
7. Good shielding practices should be followed when installing and wiring shipboard computers, communication equipment, and electronic control equipment. None of this equipment should be located in the SCR/generator room.
8. Cables should be armored according to good ABS practices.

Programmable Logic Controllers (PLCs)

Programmable Logic Controllers (PLCs) have existed for over more than 20 years as devices used for reducing the space, lower reliability, and maintenance of relay logic. Programmable controllers contain solid-state logic devices to make logical decisions and more recently have incorporated microprocessors for more sophisticated control decisions.

The basic controller accepts signals from input devices such as switches, pushbuttons, relays, and other devices and produces an operating command to other equipment. It also accepts feedback from sensors, level alarms, and other devices to maintain control of a process.

Modern PLCs have matured significantly in their hardware capability, and the programmable microprocessor offers a wide range of flexibility in software options. Today, approximately 84 manufacturers offer 365 models of PLC hardware, but the major decisions to be made by designers and application personnel involve the software and programming.

Early PLCs used Relay Ladder Logic (RLL) diagrams to guide the programmer in his PLC design in easily understood language. Electricians could follow a common method and implement the desired logic control for relatively simple applications.

Now, with years of experience behind it coupled with new technology used to expand its capability, the PLC has developed into a relatively sophisticated unit. The challenge for manufacturers now is to utilize the PLC capability with simple programming methods for the user.

The most commonly used programming language for PLCs is RLL, a form of graphic user interface used many years before the term was coined. Virtually all manufacturers use this language as part of their instructions for programming.

A less common computer language used is BASIC, and even less used is the Grafcet and Boolean languages. As the PLC looks more and more like a computer, the more its programming language will resemble computer terminology.

Petroleum Industry Applications for PLCs

Early applications of the PLC appeared on SCR-type rigs in the late 1970s since motor assignment and safety interlocks required a significant number of control relays. Using the PLCs in this area significantly reduced the number of relays used and resulted in a decrease of panel space requirements and wiring.

On production platforms, PLCs are being used more and more for equipment control and computer-controlled applications. Where remote operation is required with central land-based computer control centers, the PLC offers an excellent interface for platform electrical equipment.

Programmable Controller Applications

Programmable controllers are the solid-state equivalents of relay controls. They are housed in modular assemblies, are smaller and take up less space, and require a minimum of wiring.

Programmable controllers work in much the same way as relay logic, but much of the labor required to create the relay logic and wire it is eliminated. To make the transition in understanding the programmable controller, new terms must be learned and understood. The following explanations will serve as a basic initiation into programmable controller operations. However, once these basics are learned, the student should read a specific programmable controller manual to become more familiar with its hardware.

Some programmable controllers display the relay equivalents on the surface of the unit in graphic drawings and status lamps. This was a feature of earlier units to provide a transition from relay logic to the programmable controllers. However, newer units do not display these graphics because they can perform many flexible functions that could not be portrayed easily with graphics.

PLC Subsystems

Each programmable controller is made up of the following components:

- Processor
- Input/Output (I/O) Structure
- I/O Modules
- Industrial Programming Terminal
- Auxiliary Power Supplies

Processor

The *processor* is the central processing unit of the system. Input module signals are transmitted to the processor which operates on these data in accordance with the user's program. Then it transmits the data to the output modules.

The processor module's primary function is to provide control functions, timing, and interfacing with other processor modules. Once inserted and properly seated into the backplane socket, it is covered with a front panel which is secured to the processor chassis. In addition to covering the processor module, it also covers the memory module.

The *processor fault indicator* is often helpful in isolating processor malfunctions. It illuminates when the logic controlling the processor scan fails. This fault is usually corrected by replacing this module. The indicator can be seen through the window on the front panel cover and is labeled on the window.

The *memory module* is a read/write solid-state Random Access Memory (RAM) and contains support circuitry needed to store and retrieve memory information. It also includes a battery backup for memory backup support.

Memory fault diagnostic indicators are located on the memory module to provide lamp indication of backup battery status and memory fault condition.

The *processor interface module* contains the circuitry used for communication between the processor, user inputs/outputs, and the industrial terminal. On the front panel of this module are two sockets, two indicators and a four-position mode select switch.

The *system power supply module* converts input voltages into the proper DC voltages for the processor and I/O chassis. A single connector extends from the rear of this module and plugs into a socket located on the processor chassis backplane. This connector provides the power link from the power supply to the other processor modules and is capable of supplying current to power logic circuitry in the I/O modules.

Other functions in the processor chassis include the input fuse, I/O chassis power sockets, terminal strip and DC power protection.

I/O Structure

The *I/O structure* contains one I/O chassis, one I/O adapter module and I/O modules as required by the application.

The *I/O chassis* is a single compact enclosure for the I/O adapter module and I/O modules that make up the I/O structure.

The *I/O adapter module* provides for proper transmission of data between the local processor and I/O modules. This module is installed in each I/O chassis used with a processor and allows communication between input modules and the processor, and the processor and output modules.

Diagnostic indicators are provided on the front panel of the I/O adapter module for aid in troubleshooting.

To identify the location of a particular I/O terminal in a control system, a unique location address for programming identification is assigned. A specific designation made up of the I/O rack number, its I/O module group number and its I/O terminal number is used.

I/O Modules

Circuitry to interface various types of user input and output devices with the processor is provided by the I/O modules, of which there are three types.

Input modules monitor the on/off status of the user's input devices connected to it. *Output modules* control the on/off status of the user's output devices which in turn control the process. *Analog modules* monitor the magnitude of input and output signals which in turn are proportional to physical quantities such as temperature, weight, and velocity.

A variety of I/O modules are available for certain voltage levels, isolation, integrated or discrete device interface, switch/contact control, etc.

Industrial Programming Terminal

This unit is the primary programming terminal for the controller. It allows the maximum capabilities of the controller to be utilized, as expressed in memory and I/O points.

The terminal is used to load, edit, monitor, and troubleshoot the user's program in the processor memory. In addition to programming, the industrial terminal can be used for report generation or as an interface for the digital cassette recorder or the data cartridge recorder which can be used to record and load processor memory onto a magnetic cassette tape.

Auxiliary Power Supplies

The controller's power supply provides current to power the I/O adapter module and I/O modules. When the total current required to power all of these modules reaches its limit, an *auxiliary power supply* or supplies are needed.

The output current to each I/O chassis must be calculated to ensure that its total current draw will not exceed the rating of the auxiliary power supply to be used.

In most systems, the programmable controller will not have the capacity to handle all of the logic functions required. If this is not the case, additional programmable controllers can be added and interlocked with the other programmable controllers to form a complete logic system.

BIBLIOGRAPHY

Gutzwiler, F. W. *General Electric SCR Manual,* 4th & 5th Eds. Syracuse, New York: 1967, 1977.

Jaeschke, Ralph L. *Controlling Power Transmission Systems,* Cleveland Penton/IPC, Inc., 1978

Laduzinsky, Alan J. *Control Engineering,* "PLCs Develop New Hardware and Sofware Personalities", February 1990, pp. 53–54.

Morgan, Max J. *Dynamic Positioning of Offshore Vessels,* Tulsa: Petroleum Publishing Co., 1978.

8 ELECTRIC SYSTEM STARTUP

Initial system startup determines how well the complete electrical system performs. All the effort of the rig engineer, suppliers, and the rig construction personnel culminate when the diesel-generators are started and the rig checkout begins.

The complete drilling rig consists of the products of many manufacturers, not just those of the rig-up yard or supply store. These subsystems must work well together under all conditions of rig operation prior to delivery to the drilling site. Setting up the system in the rig-up yard and operating it under loaded conditions ensures the adequacy of the diesels, generators, SCR controls, motor controls, and all other rig functions. If problems develop at this point, availability of technical assistance and materials allows quick solutions. If these problems occur on the drillsite, the subsequent downtime, loss of productivity, and repair costs would be several times those in the rig-up yard.

The System Electric Diagram

The key element in understanding how the system functions is the system electric diagram prepared by the rig designer or the designer's associates. It combines all functions of the rig in a simplified, one-line format and gives immediate reference information to operating personnel. Basic functions and equipment such as the generators, lighting, auxiliary AC loads, and the SCR drives are shown in this diagram (Fig. 8–1). Supplementing the system diagram are individual wiring diagrams and schematics of individual subsystems and controls, usually supplied by the manufacturer.

To avoid or minimize problems at startup, the rig designer/manager must plan and coordinate all activities with suppliers. While each supplier usually provides quality, well-tested equipment, problems may occur at the interface of equipment supplied by different manufacturers. Often, responsibility for these interface problems is placed with the manufacturer; however, responsibility should rest with the rig designer/manager. The rig designer/manufacturer is the one individual with more knowledge of the system operation than anyone and should anticipate and resolve interface problems before they occur.

A checklist of typical areas where interface problems may occur includes:

- Voltage regulator to generator
- Governor controls/actuator to engine
- Generator circuit breaker to generator
- Motor control center to rig motor loads
- Lighting panel to rig lighting loads
- Plug panel of SCR equipment to loads
- Cable size/type to interconnections
- Engine-generator capacity to all loads
- Driller's console functions
- Applicable standards and certification requirements such as ABS, USCG, API, etc.

Cable and Wire Selection

The *National Electrical Code* (NEC) establishes the minimum wiring standards in the United States and should be consulted for all designs. Other standards which may exist in certain applications and should be checked include: ANSI, USCG, ABS, Lloyds, Det Norske Veritas, state and local codes, and OSHA.

The rig builder uses an internal electrical contractor or hires one to prepare the information required to select the cable and wiring for the rig. Suppliers of the AC generator, lighting, SCR equipment, generator controls, mud pumps, drawworks, etc., define the current and voltage or motor size. The rig engineer and the electrical contractor must properly interconnect this equipment.

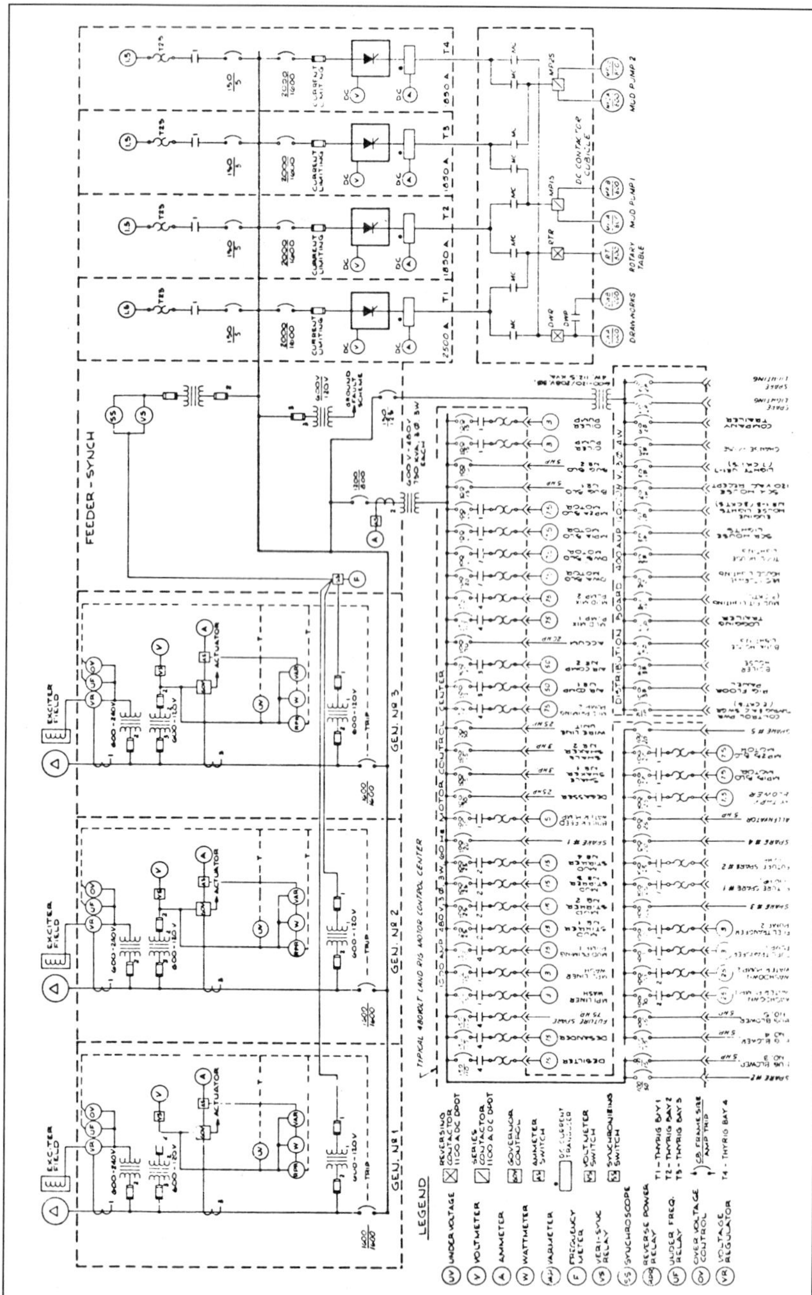

Fig. 8–1 Typical one-line electrical rig diagram

During the wire and cable selection, one must consider equipment current and voltage ratings to determine the conductor size and insulation ratings, length of cable required throughout the rig, its voltage drop, and the method of securing it over its routing. In addition, environmental factors, such as where the rig will be operated (e.g., hot or cold temperatures, high or low humidity, fungus levels) must be taken into account. Flexibility or rigidity requirements of the cable, depending on its use, and the need for frequent rig moves resulting in frequent coiling and uncoiling of cables must also be considered. Finally, specifications that require special insulation or protection, such as armored cables under USCG inspection, and cable runs to conduits or open trays that require different conductor sizes and insulation due to temperature variations must be noted.

In general, most rig cabling is for applications of 600 volts or less and can be run in the same enclosure. Cables carrying voltages more than 600 should never be mixed with lower voltage cables and should be in separate distribution ducts. Communication or low-level signal cables should be run in separate conduits or cable trays to reduce electromagnetic coupling of power noise.

Conductors should be sized for at least 125% of the full-load rating of each motor when individual conductors are provided. When multiple motors are operated from a feeder circuit, the conductor should be sized to provide current for 125% of full-load rating of the largest motor plus the sum of the full-load currents of the remaining motors. If the maximum duty cycle is known, this conductor size may be reduced with permission from the regulating groups governing the installation.

The voltage drop from the power source to any motor should not exceed 5% at a power factor of 0.80. Calculations of this drop should also incorporate resistance and reactance of the circuit.

Engine-Generator Startup

Prior to startup, all operating personnel must be trained and briefed on the events to take place. Check engine-generator couplings and alignment, mounting, fuel lines, fuel actuator movement, speed sensor, and electrical connections to the generator switchgear. Thorough knowledge of the controls is important before starting the generating system.

Start the engine manually and allow an adequate warm-up period before operating the generator at rated voltage and speed. Once rated speed is reached, adjust the generator controls manually until the proper voltage is achieved. Switch generator to automatic control of voltage and frequency and adjust for correct values. Close the first generator circuit breaker to connect its output to the system bus. Bring the next engine-generator up to proper speed, voltage, and frequency. After bringing the second unit into synchronism with the system bus, parallel this generator to the bus. Fine-tune the voltage and engine speed to reduce circulating reactive and real components of current from the generator. Repeat the sequence with other engine-generators.

Once the engine-generator system is operating in parallel and properly load sharing, any generator can be removed from the bus with the remaining units continuing to supply power.

Switchgear and Motor Control Center

After the system bus is energized, check all circuit breakers feeding power transformers, SCR units, and other loads. Until this load equipment is started separately, little current should flow from the generators. Operate each starter and circuit breaker sequentially on the motor control center. Check operation and direction of rotation of each AC motor on the rig. Correct any operational faults or other errors before proceeding with the SCR controls.

Rig Lighting, Hotel Load, and Auxiliary Functions

Test all rig lights from the lighting panel and correct any defective bulbs or circuits. Test hotel load functions (fans, air conditioners, ranges, etc.) for proper functioning.

Mud Pump Operation

Prior to energizing the SCR mud pump bay, set all driller controls to "off" and place all assignment selections in the proper position according to manufacturer's instructions. Energize the mud pump SCR bay and proceed to operate it over its range. Correct any faults that may occur during this sequence.

Drawworks Operation

Remove power from the mud pump SCR bay and energize the drawworks SCR bay. Operate the drawworks in forward and reverse direction and test other functions such as dynamic braking. Correct any faults or errors as they are discovered.

Rotary Table/Top Drive

Remove power from the drawworks SCR bay and operate the rotary table or top drive. Check all functions including reversing and correct faults as they are found. Change assignments of loads of SCR bays and repeat the checkout. Operate all SCR units together at no load and repeat with load on the drives.

Driller's Console

Operate all emergency functions and test all instruments for performance and accuracy.

Thruster Operation

If the system is an offshore rig, test the thrusters under all modes of operation and correct any faults as they occur.

9 RIG AUTOMATION

Introduction

Drilling rigs have evolved over the past 50 years from slow-moving mechanical systems driven by internal combustion engines, to rotating electrical generator-motor systems, to the present electronically controlled, SCR variable-speed drives. This evolution has created electronic tools and downhole data acquisition devices that reduce the guesswork involved in the drilling process.

While significant improvements have been made in the drilling industry, the technological changes lag considerably when compared with other segments of the petroleum industry.

For example, the production and processing segments of the petroleum industry have introduced microprocessor-controlled data acquisition and control systems for offshore production platforms, pipelines, and petrochemical processing plants. Supervisory Control and Data Acquisition (SCADA) systems use automatic computer controls to monitor and control offshore platforms from a central land-based location as shown in Figure 9–1. In these areas, computer control has been applied to reduce manual labor, optimize processes and performance and improve safety in operations.

Throughout petrochemical plants, we find variable-frequency, variable-speed AC motor drives in pumping and mixing applications as shown in Figure 9–2. Pipelines use these same variable-speed drives at booster stations to pump liquids at a reliable, remotely controlled pace with reduced maintenance.

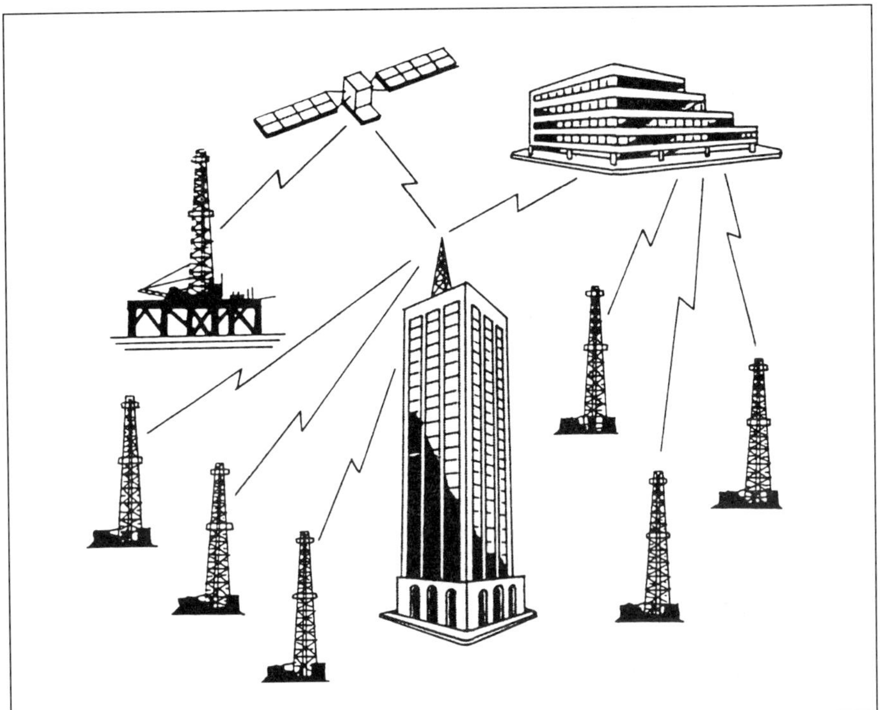

Fig. 9–1 Typical Automated SCADA System

The conclusion we draw from witnessing these applications of technology is that the drilling industry has an opportunity to solve some of its current and future problems with this technology.

Drilling Industry Needs

The legacies of the 1980s drilling industry downturn and depression include the stacking of land rigs and offshore MODUs, very low sales of new rig equipment except in occasional areas outside of the United States, low investment in new technology and rig improvements, significant layoffs of personnel, closing and merging of drilling companies, and an attrition of experienced "hands" from the industry.

The drilling industry of the early '90s is one that has reorganized, refinanced, restructured, and refined operations to the point that a skeletal structure now exists. There are signs of a slight upturn in drilling, and the first requirement is to hire more "hands" to get the rigs back to work. The problem with adding new personnel, however, is that the reservoir of qualified personnel has dried up since the industry no longer needed them during the period of low productivity.

Herein lies an opportunity for the drilling industry to identify its future direction by utilizing technology that supplements its need for personnel.

Establishing Technical Direction

The drilling industry's methods of operation have historically been defined by the needs of the producer and operator. These needs can be defined in terms

Fig. 9–2 Variable-speed, variable-frequency drives (Courtesy Reliance Electric)

of minimum lost time, minimum lost holes and fishing jobs, maximum safety to personnel, effective drilling rates and the drilling costs. The successful drilling contractor bidder was the one who could produce the rig and personnel that the producer needed.

If we view what the producer/operator has done for production, refining, and pipeline operations, we see a picture of high technology, an emphasis on safety and the methods of optimizing performance while reducing cost. If the drilling industry follows these guidelines, it will find itself in a very strategic position to meet the current and future needs of the industry.

Current Drilling Technology

What we have in the industry, despite the '80s downturn, is a modern drilling rig that offers these technology improvements:

- SCR-type variable-speed DC motor drives
- Top motor drives that reduce stuck bits and lost holes
- Measurement-while-drilling systems that provide downhole data for improved performance
- Directional and horizontal drilling techniques that improve downhole production
- Computer-controlled automatic weight-on-bit coring controls that reduce drill bit wear and increase core lengths.
- Automatic pipe handling systems that lift, position, and makeup/breakout drillpipe
- Electronic-controlled BOP systems

Each of these technology improvements meet the needs of the producer/operator, which include:

- Improved safety
- Lost hole reduction
- Reduced drilling time
- Improved drilling performance

In most cases, however, the drilling contractor's rate of compensation will increase when new technology is added to the equipment. But the producer/

operator is willing to pay this premium in exchange for the boost of confidence added benefits produce. In these cases, the cost represents a win- win situation for both parties.

Figure 9–3 shows a typical top drive system that allows the driller to drill complete stands of pipe (three lengths) before connections are made, as

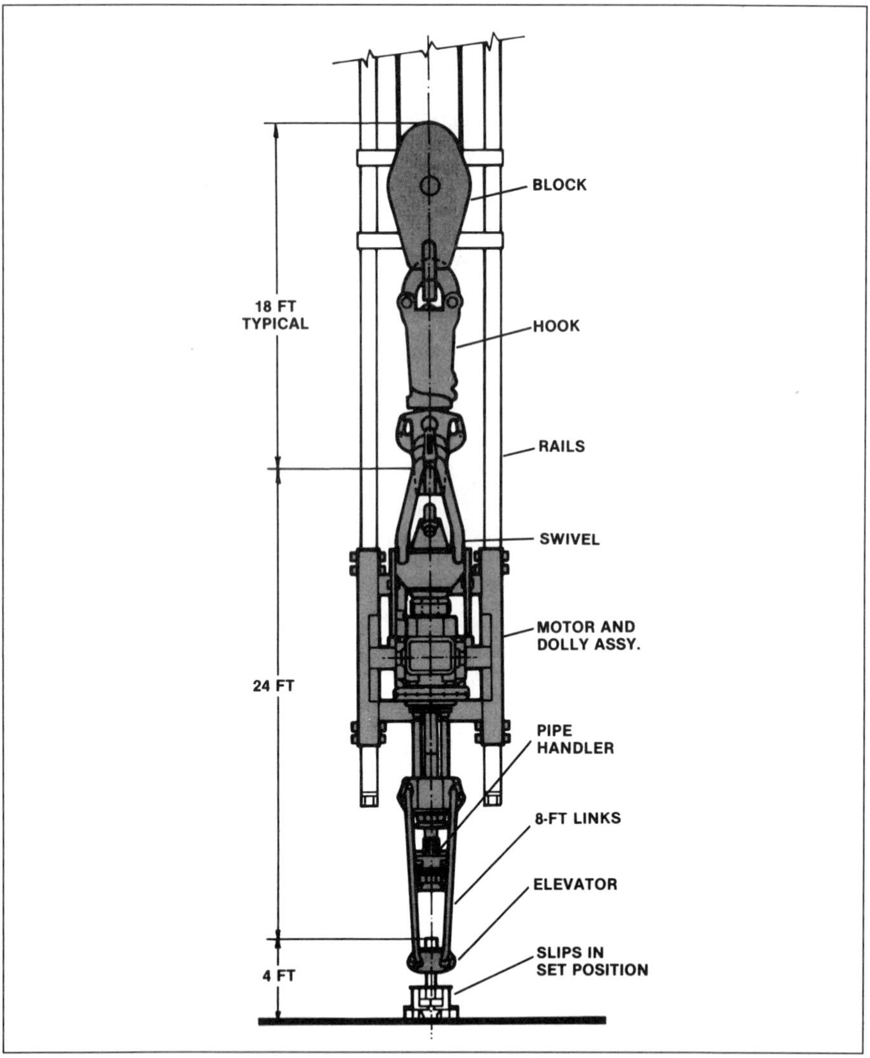

Fig. 9–3 Typical top drive system (Courtesy Varco International Corp.)

compared to individual lengths of drillpipe used in rotary table-operated systems. This reduces lost time that can be used for productive drilling. Another significant advantage of the top drive is its ability to rotate while the bit is lifted off bottom, thereby reducing the possibility of a stuck bit and resulting fishing job.

Significant advancement also has been made in coring, with the automatic weight-on-bit control system introduced by Parker Drilling Company. The system uses an electronic load cell positioned above an in-line top drive motor to weigh the complete drillstring, as shown in Figure 9–4. Once the bit touches bottom, the drillstring relaxes an amount equal to the weight on bit. Using this form of feedback through a closed-loop automatic control system, the bit weight is maintained within 0.1% of the desired set point. After each core section is drilled, new data update the computer to continue the coring process.

Automatic pipe handling equipment has been introduced on a limited basis on small land rigs and on European offshore rigs. On a long-term basis, this

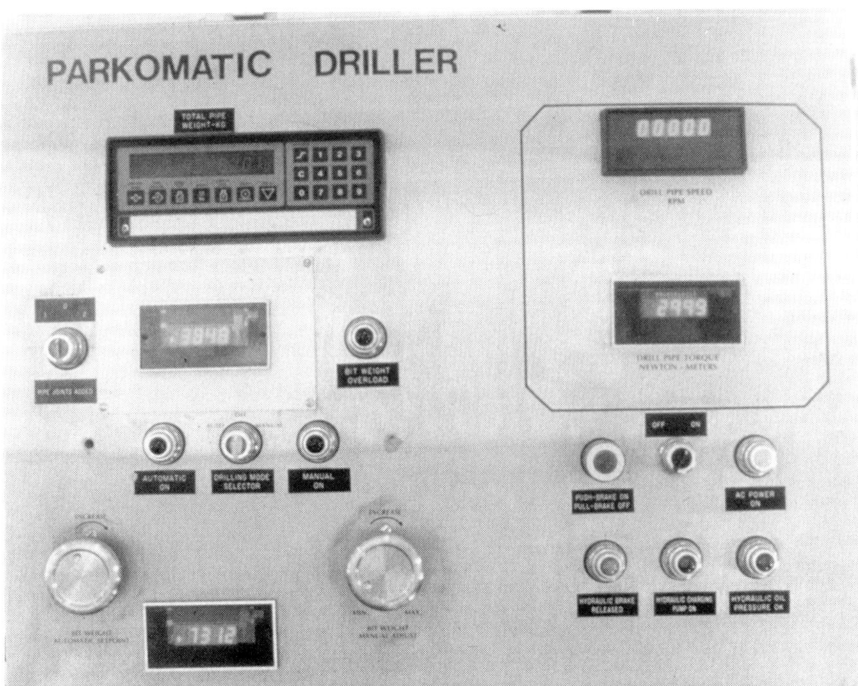

Fig. 9–4 Automatic electronic weight-on-bit control (Courtesy Parker Drilling Co.)

development represents a major achievement in improving safety for personnel and reducing damaged pipe threads caused by improper positioning. A typical system is shown in Figure 9–5.

Downhole drilling techniques are allowing drillers to accurately direct their bits diagonally or horizontally to penetrate multiple pay zones with the same drilling operation. Significant production improvements are being made using these techniques which involve sophisticated electronic packages downhole and at the surface.

Power management controls for the diesel/generator packages have been available for many years, but few drilling contractors have taken advantage of them. Complete application of this technology would allow rotation of individual units to reduce maintenance, automatic startup and removal when power demands require this action as well as proper loading to optimize use of diesel fuel.

Defining Automation

The evidence we see in the technology being developed and added to new rigs is a trend toward automatic rig operation. While individual pieces of equipment are currently being employed to perform certain functions automatically, these can form the basis for future drilling automation.

Automation has been employed in many industries for the following reasons:

- To replace manual, repetitive tasks with machines
- To remove personnel from hazardous areas
- To improve processes by more precise controll
- To produce better quality at higher performance levels
- To optimize the time to perform certain functions
- To perform multiple functions simultaneously

If we analyze each of these reasons, we find many applications in the drilling process. For example:

1. Setting up a drill rig at spud-in could be accomplished with automatic machinery, reducing the risk of exposure of personnel to injury.
2. Handling and processing the drillpipe could be accomplished with automatic pipe handling equipment with minimal manual effort required.

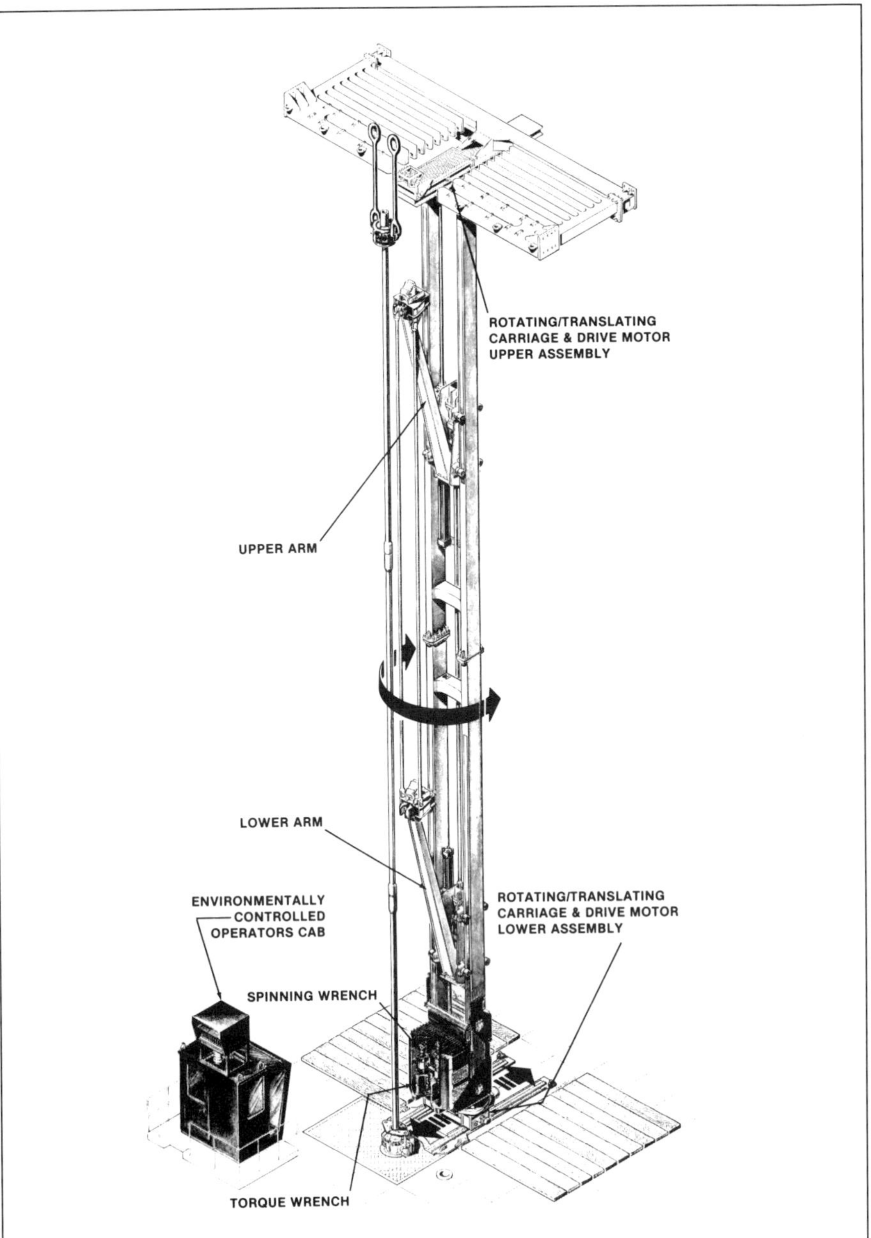

Fig. 9–5 Automatic pipe handling system (Courtesy Varco International, Inc.)

3. Automatic drilling controls to maintain bit weight, penetration rate or other variables could be used to improve the drilling process.
4. The mud system could employ mud mixing and weight control, flow control, and pressure detection.
5. Variable-speed drives could be controlled from a central console to optimize drill bit speeds based on downhole conditions.
6. Diesel/generator sets could be automatically started, paralleled, load-shared, and removed based on operating conditions of drilling or tripping.
7. Detection of downhole gas "kicks" signifying potential blowouts could be automatically controlled with increased mud weight or shut-in of the BOP system as required.

Automation Equipment and Components

The obvious question to ask when considering automation of drilling rigs is: "What type of equipment must be added to existing rig designs to automate the process?" While this question cannot be answered in simplistic terms, a general grouping of component functions can be identified. These functions include the Power System, Actuator, Sensors, Control Signal Processing and Control Cabling.

Power System

In the early days of drilling, mechanical power from diesel engines was used through chain and gear transmissions to perform the functions of rotary drilling, mud-pump control and drawworks hoisting. From that point, hydraulic and pneumatic power techniques were implemented, primarily because of the hazardous gas presence on the rig. As the rig has evolved to its current design, more and more electrical power equipment has been added, with explosion proof types being used in hazardous areas. While some hydraulic and pneumatic equipment will still find suitable application on the rig, the rig's primary source of energy will be electrical power.

Actuation

The component used to move, position, accelerate, torque, and change equipment such as pipe- and mud-handling systems is the actuator. Actuating equipment can be powered by either electric, hydraulic, or pneumatic means to produce the desired mechanical output of torque, pressure, force, speed, and flow. Since a significant amount of automation is based on electronic controls and sensors with electrical outputs, the obvious and simplest method of interfacing is to electrically powered actuators such as with electric motors and linear actuators.

Sensors

Devices used to sense equipment position, speed, torque, pressure, flow, depth, tension, strokes, temperature, acceleration, and other physical parameters are referred to as sensors. The most reliable and accurate types currently in use in the industry are the ones using some form of semiconductor element that produces an electrical output. It is this type of sensor that interfaces well with electronic signal-conditioning devices and in closed-loop control systems.

Control Signal Processing

With the introduction of the personal computer and other digital devices, the device that has had the most significant impact on its acceptance has been the microprocessor. It has the capability of accepting multiple inputs and producing an output in response to a predetermined program description. The microprocessor is the ideal device to be used in the design of an automated rig in a closed-loop control system.

Control Cable

One of the critical elements involved in a control system is the electrical cable used to transmit low-power control signals from the sensor to the controller and to the actuating device. Small copper cables can be used in some cases, but due to the significance of the signals being transmitted, shielded pair cables are preferred. Both analog and digital signals can be transmitted over these cables. With proven applications, fiber optic cables have become the preferred means of transmitting digital signals in an environment containing "noisy" electrical power such as that which exists on drilling rigs.

Automating Existing Rigs

While most of us have envisioned an automatic rig from the standpoint of a specially designed machine that is effectively out of reach for most contractors, limited automation can be implemented into existing rigs. By using a modular approach where each major drilling function is packaged separately into an automated package, varying degrees of automation can be achieved to meet particular needs. Typical segments of the rig that can be modularized include:

- Pipe-handling system to move pipe from ground level to rig floor prior to makeup connections;
- Pipe makeup and breakout system, including positioning and torquing controls;
- Power management of the diesel/generator sets;
- Mud system control which would include mixing mud chemical, water/oil/chemical additives, mud weight control, mud flow control, mud loss detection, mud gas detection, shale shaker and filtration system; mud temperatures, salinities and alkalinities;
- Rotary drilling controls of bit speed, bit weight, penetration rate, direction of rotation, torque, and drawworks lifting controls;
- Tripping controls which would include the drawworks, drawworks braking, and pipe inventory controls;
- Hole safety controls, including gas and pressure detection and BOP operation;
- Use of AC variable-frequency drives using AC motors as compared to current use of SCR-DC motor drives to reduce maintenance and improve safety;
- For systems using directional and horizontal drilling equipment, these can be added to the system as a separate module and integrated into the total system operation;
- Data acquisition of many drilling variables, including downhole pressure and formation information, bit weight, bit speed and penetration rates, mud flow and pressure information, hole depth, and gas detection.

One or any number of these modules could be added to existing rigs with some design and planning effort. The power modules would perform the actuating functions while the control modules would be incorporated in a central console.

With a system of this type, personnel could be removed from the rig floor during normal operations, while considerable activity could be accomplished from the central console by the drilling operator.

The Modular System

Figure 9–6 shows a typical system diagram of anautomated rig system with its functional blocks. Individual blocks could be incorporated into existing or new rigs or the complete system could be adapted for best results.

This diagram also shows two paths of control: One is the power bus and the other is the control bus, where commands are given and sensor information is

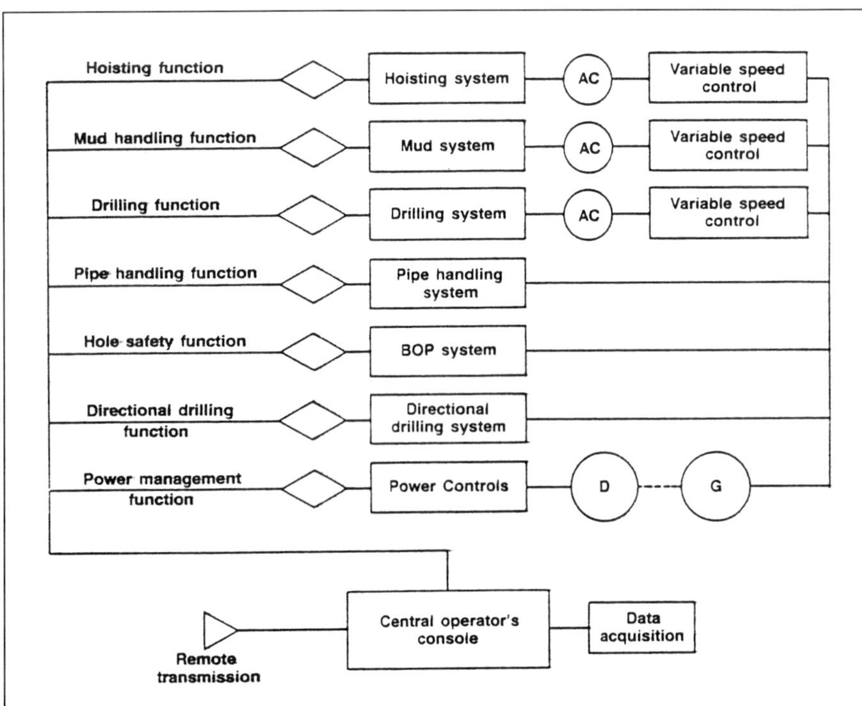

Fig. 9–6 Block diagram of an automatic rig system

received by the central microprocessor in the master console. Use of this concept would allow standard control signals to be implemented so that compatibility between different manufacturers could be established in much the same way that personal computers use standardized bus architecture. A standardized method of operation as described by an algorithm would form the basis for a modularized, automated rig.

A Current Application

Adding sensors and data acquisition devices to existing or new rigs is a practical solution today. Typical of this is the system developed by one drilling contractor as described below.

Rig Data Acquisition

Gaining an understanding of the details of drilling requires sensors and instruments at the point under consideration. On most drilling rigs, there is very little information available to the driller since most rigs have a limited amount of instrumentation and mechanical gauges.

Some of the drilling parameters that could be measured and would aid the drilling process include:

- Hook load
- Pump pressure
- Mud pit levels
- Rotational speed of rotary or top drive
- Torque of rotary or top drive
- Pump strokes
- Standpipe pressure
- Return line paddle deflection
- Mud-tank and trip-tank volumes
- Heave on floating rigs

From this information, a number of calculations could be made that would give the driller and rig crew even more information, including:

- Depth of bit

- •Electronic tally of pipe depth
- •Weight on bit
- •String speed
- •Rate of penetration
- •Flow into and out of well

Even further, data acquisition would allow high- and low-level set points to be established to sound alarms or limit conditions of operation. For example, when tripping, string speed is monitored. If the speed goes too high, risking swabbing or surging, an alarm could be sounded. When drilling, determining the drillpipe torque to avoid a twist-off would be beneficial with a high-torque limit. Monitoring and alarming a low mud pit level would be beneficial to determine formation losses.

If a rig were to be automated, these signals could not only be used to provide information and perform calculations, they also could be used as part of a "closed-loop" control system to provide feedback information.

At least one international drilling company has developed a system that

Fig. 9–7 Operator's view of data acquistion informationusing the Sedco-Forex MDS system (Courtesy Sedco-Forex Schlumberger)

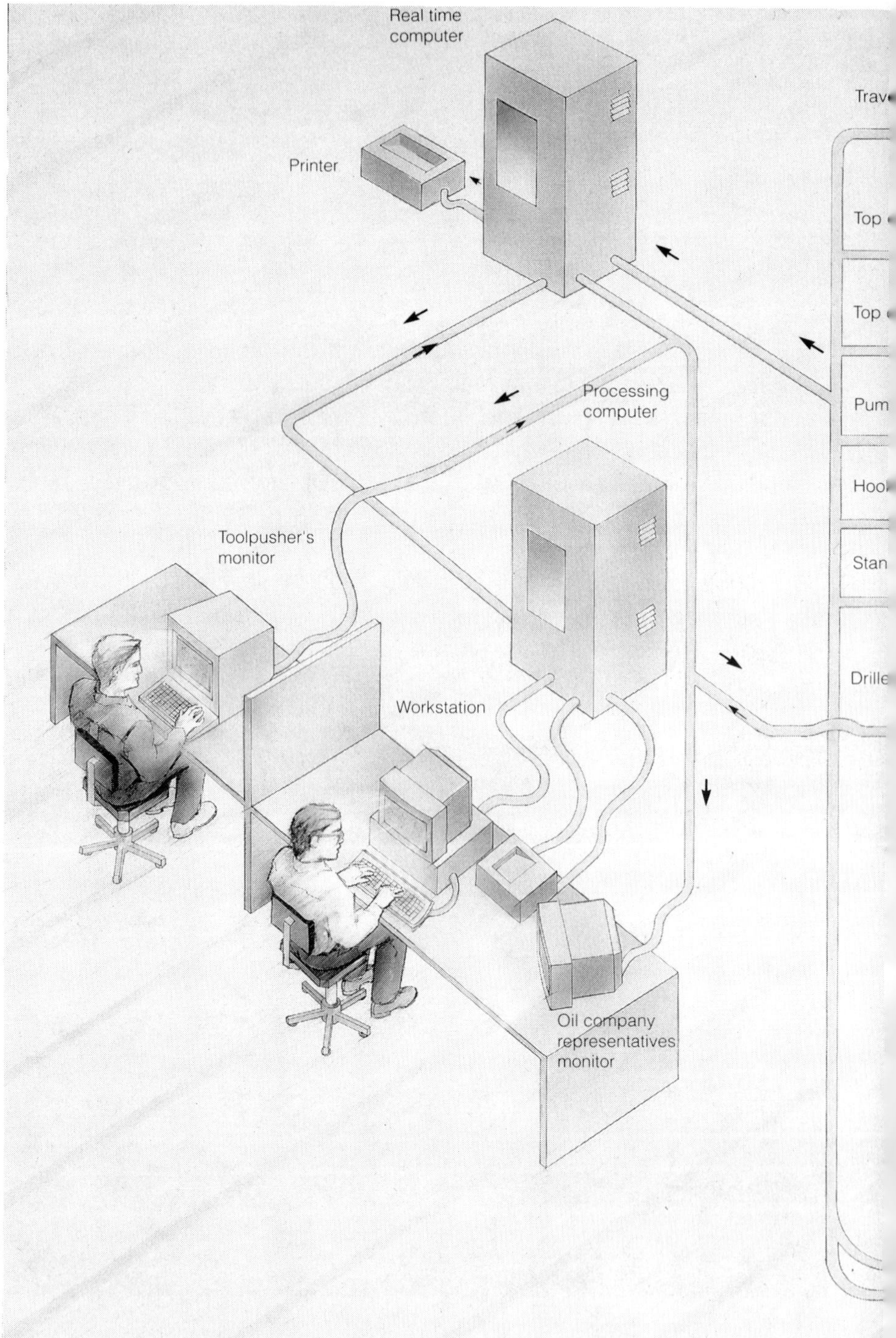

Fig. 9–8 Simplified diagram of MDS system application (Courtesy Oilfield Review)

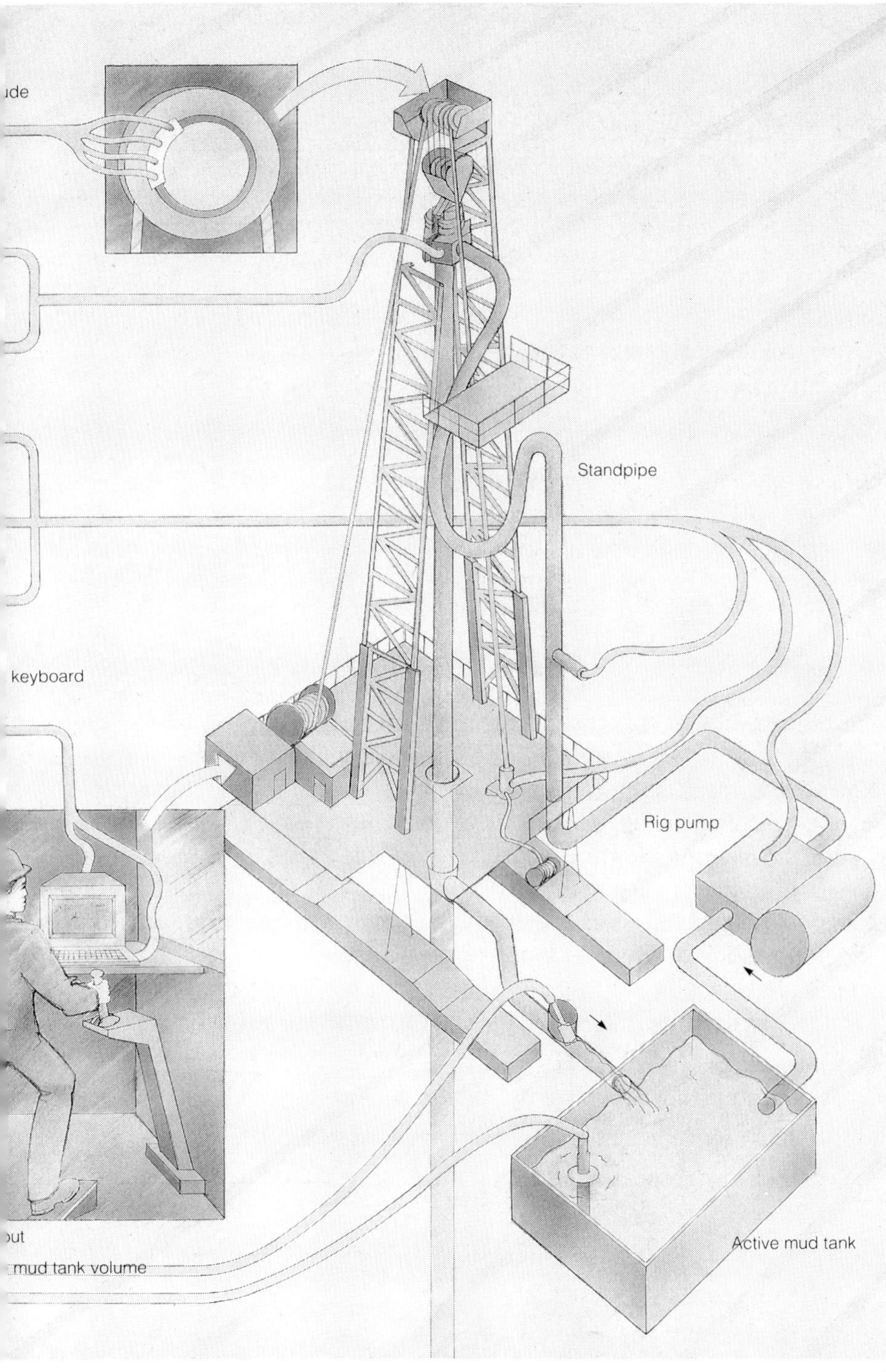
ıde
keyboard
Standpipe
Rig pump
Active mud tank
ɔut
mud tank volume

provides such a monitoring and data acquistion system using a computer and display. The operator's view of this system is shown in Figure 9–7.

A typical application of the system using the Modular Data System (MDS) system is shown in the simplified diagram of Figure 9–8. This system uses two computers: one for real-time data acquisition for the driller and the other to format data for access by the oil company representative and the rig superintendent.

Three color displays are provided: One on the rig floor and one each in the rig superintendent's and oil company representative's office. These display real-time drilling data.

Sensors used with the system may be either standard off-the-shelf type or custom-engineered for specific applications. Typical of these sensors are displacement sensors for measuring length and motion of pipe, load cells for weight measuring, flow meters for mud and liquid flow rates, pressure sensors for mud pump and downhole pressures, and others.

The Auto-Rig

In the early 1980s, an attempt was made to design an automatic drilling system, but significant technical and operational problems could not be overcome economically. Since that time, there have been significant advances in technology that make the *Auto-Rig* a real possibility. One of the essential elements is the microprocessor, which allows multiple inputs to be processed to a desired, programmed output for control. Using the microprocessor as an integral part of the system design, techniques such as multiplexing and fiber-optic transmission of signals in a noisy rig environment can be economically implemented.

Some considerations that would have to be made in the design of the Auto-Rig for oilfield use would include:

- Redundancy in the control signals and cabling
- Easy implementation of replacement modules
- Self-diagnosing to reduce the need for high-level technical assistance
- Manual overrides as required
- Ruggedized construction
- Simplicity of operation

Planning For The Future

If proper action is taken by the drilling industry, a significant change can be made in the character of its operations. By continuing to apply new technology to reduce the activity of floor hands, we will see the employment of more technically trained personnel. Instead of "roughnecks," we will have "rig-techs;" instead of "drillers," we will have "operators." Technicians will replace electricians. And the term "hands" for personnel may become obsolete with the reduction in manual labor.

The end result of the Auto-Rig will be a system that allows machines to be hurt, not personnel; a drilling system whose performance can be optimized through data acquisition; a system that becomes compatible with ever-advancing downhole drilling technology; and a system that utilizes more technically knowledgeable personnel as compared to those who learned their art through experience.

Automating and data acquisition has a number of side benefits beyond the actual field site activities. One of these is the ability to compile drilling progress information (the "morning report") in a computer and transmit it via telephone or satellite communication on a real-time basis to the home office. In this way, the rig superintendent, who generally oversees several rigs, can observe each rig's activity, problems and performance from a desktop computer. Immediate direction can be provided via the computer link to the field when needed and office technical expertise can be summoned to solve rig-related problems.

Drilling Automation Benefits

The major benefits of drilling automation include:

Safety: Instead of achieving 9.17 accidents per million manhours or even 3.29 accidents per million manhours as quoted by other contractors, automation has the potential to approach *zero* accidents on the rig floor and around pipe operations by eliminating personnel from these locations.

Improved performance: Personnel are limited to how fast and effective a job they can perform while automated equipment operations can be optimized through control of position, speed, and acceleration and by performing multiple functions in synchronism.

Reduction in lost holes: Use of top drives that allow circulation and rotation while pulling off bottom in conjunction with mud sensors aids in preventing blowouts and stuck bits;

Reduced drilling time: The driller's time at the drillsite determines the cost of creating a hole whether or not the hole produces. Automation reduces the time required for transporting equipment, setting up, drilling, tripping, and other operations.

Approaches To Drilling Automation

One of the major concerns of drilling contractors toward automation is their current investment in rig equipment. Many have invested millions of dollars to obtain their current rig inventory and want to make the best use of that equipment without further investments.

Drilling contractors are faced with several issues that developed from the industry's downturn in the 1980s. Among these are the current lack of trained rig personnel who left the industry for other occupations, the move by other working drilling contractors to upgrade their rigs with new technology such as top drives, and the inertia of maintaining the status quo of using existing methods to drill for oil and gas.

Drilling contractors who decide to utilize automation technology may choose one of two approaches. The first approach is to use existing rigs and make incremental changes toward rig automation, such as adding a top drive, then automatic pipe-handling equipment, and so forth. This approach keeps major investments intact while adding improvements that are attractive to the producers.

The second approach is to purchase a completely automated rig that incorporates all of the features that provide the desired safety, performance, and reliability of operation. This rig would include a single console operated by one "rig-tech" and supported by one or two maintenance/support personnel. It would include automatic pipe-handling equipment from the ground level to the makeup/breakout operations; sensors to detect well and equipment conditions; actuators to perform mechanical activity such as rotary drilling, mud-pump operation, mud-tank mixing and cleaning operations; and safety controls to detect and shut down systems in the event of hydrogen sulfide detection, gas kick, or mechanical equipment failure.

A rig designed specifically with complete automation usually is more cost-effective, smaller, lighter, and easier to operate than one with incremental modifications.

Personnel Considerations

Implementing automation in drilling operations requires a review of the type of personnel required for operating the rig system.

Depending on the level of automation, the rig crew still will require an experienced driller who understands the techniques of hole construction, mud composition, downhole conditions, casing, and completion. Also required is an individual who has technical knowledge of the system operation and equipment functions, who can analyze problems associated with the system and has the ability to maintain operation in the event of equipment failure.

Other support personnel may be required depending on the size of the rig and whether or not it is land- or offshore-based.

The automation crew should be capable, experienced drilling personnel with the technical ability to handle operations and problems associated with electromechanical equipment.

Automation usually eliminates personnel who are required to handle pipe or casing, makeup and breakout connections; to use spinning chains, slips and tongs; and to provide derrickman and floorhand functions and other operations that are considered to be dangerous. Those operating the rig are located remotely from moving mechanical equipment and away from overhead equipment.

Operating Major Drilling Functions

By breaking down the drilling rig operations into job functions, we can analyze how the operator will perform the various jobs on the automated rig. Among these jobs are:

- Pipe handling from ground level to the rig floor
- Making connections properly and torquing threaded joints
- Controlling the drillstring with restraining slips
- Lifting and lowering the drillstring
- Turning the drillstring and bit for drilling and coring
- Operating the mud system
- Tripping out of the hole
- Breaking out connections
- Removing pipe in tripping out operations

- •BOP operations and control
- •Casing and completion operations
- •Testing and logging operations
- •Rig-up and rig-down operations

Many of these operations use the same equipment to perform the functions required and usually operate in synchronism with related equipment.

Pipe-Handling Equipment

To utilize automation fully, the pipe should never be touched or handled by rig personnel. The pipe should be unloaded onto a specially designed pipe rack that allows the individual units to roll into position one by one for processing into the derrick.The horizontal pipe rack is typically slanted to roll the individual pipe into the remote-controlled pipe-handling mechanism as shown in Figure 9–9.

After entering the pipe-handling mechanism, the pipe is moved into position above the existing drillstring and automatically screwed and torqued with the equipment shown in Figure 9–10 before being released.

During this operation of pipe makeup, the drillstring is automatically restrained in automatic slips on the drill floor. After the connection is made, the drillstring is slightly lifted, slips released, and drilling is resumed with the rotary table or top drive under controlled conditions.

Tripping out operations follow a similar pattern: The drillstring is restrained and held in the automatic slips, automatic wrenches unscrew the tool joint, and the pipe handling mechanism removes the pipe from the derrick and places it on the horizontal pipe rack for storage.

Drilling Operations

Automated drilling operations use several methods of control. First, the control of drill bit speed is determined by electronic variable-speed DC or AC drives that can be adjusted to precise settings. On an automatic rig, the speed is programmed into the drilling program and is automatically resumed after a connection is made. In the event of drillstring harmonics or vibration at a specific speed, the speed can be automatically adjusted to stay away from these conditions. These controls also adjust the maximum torque that can be applied to the drillstring while turning the bit.

Fig. 9–9 Automatic pipe handling equipment (Courtesy Walker-Neer Corp.)

Secondly, bit weight as well as penetration rate can be controlled by altering the amount of drillstring weight on the bit. Several systems have been designed to accomplish this, including the one by Parker Drilling Company shown in Figure 9–11.

Fig. 9–10 Automatic pipe torquing equipment (Courtesy Varco International Corp.)

In the event mud motors are used in operations such as horizontal drilling, the motor speed and bit weight can be controlled with the use of feedback sensors and controls in the automatic system.

Pipe-length measuring techniques can be employed in the system to determine the depth of the drill bit and to compute the actual and average penetration rate over various intervals of drilling.

The method of turning the drillpipe can be selected between the traditional rotary table or the top drive. The top drive has proven its ability to continue turning the bit while pulling off bottom, reducing the possibility of a stuck bit and subsequent fishing job. The top drive system is produced by several manufacturers.

An automatic rig operator's console is shown in Figure 9–12.

Tripping Operations

Automatic tripping-in of drillpipe is accomplished with the pipe-handling mechanism, automatic slips, top drive (if used), and automatic torque wrenches. After the pipe is positioned above the drillstring secured in the slips, the pipe is lowered until the threads come into contact with each other. The top drive is used in most cases to provide the initial threading of the tool joint. Final torquing of the tool joint is accomplished with the automatic wrench assembly to the desired preset torque level.

After the connection is made, the pipe-handling mechanism is removed, the drillstring is lifted with the drawworks and the slips released. The drillpipe is

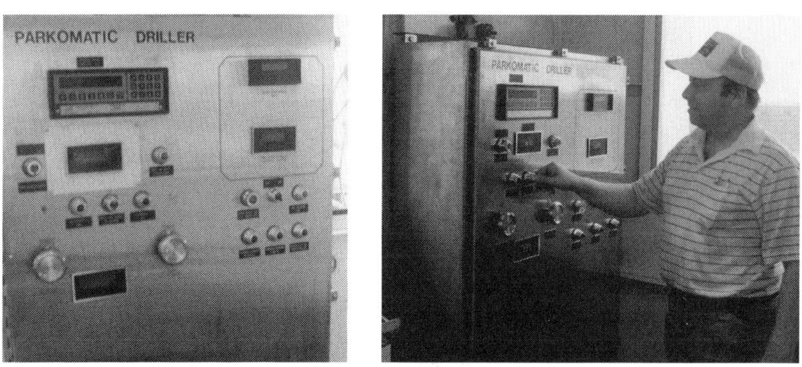

Fig. 9–11 Automatic weight-on-bit controller (Courtesy Parker Drilling Co.)

Fig. 9–12 Automatic rig driller's console (Courtesy Walker-Neer Corp.)

then automatically lowered with the bit turning until the desired bit weight is achieved.

Tripping-out operations are essentially reversed from the above description. The drawworks lifts the drillpipe out of the hole until the tool joint clears the slips. The slips are actuated and the pipe-handling mechanism moves into position to grasp the pipe section to be removed. After the torque wrench unscrews the pipe, the top drive completes the separation and allows the pipe handling mechanism to lay the pipe section down on the horizontal pipe rack.

All of these tripping operations are performed with the associated equipment moving in synchronism with one another for the optimum operation time.

Mud-Control System

The mud system uses powdered barite and a liquid additive that can be a chemical, oil, or water depending on the hole requirements. As the mud weight needs to be changed, the liquid additive will alter its density and thus its weight. The process of mixing these components offers the opportunity to use an automatic batch-mixing process to further automate the drilling operation.

Further automation can be achieved with the mud pumps by automatically controlling the system's speed and thus mud volume flow rates. Downhole pressures can be monitored to ascertain the pressure required by the mud pump and the mud weight the system requires.

Automatic BOP Controls

Further protection of the hole, equipment, and personnel can be accomplished with hole sensors detecting gas and pressure conditions. If certain limits are reached that require rig shutdown and closure of the BOP, an automatic BOP control system would work in conjunction with the operator's console to achieve this goal.

In the past, manual closure of valves was required, but the same operation can be accomplished automatically with fail-safe electropneumatic or electrohydraulic valves. Manual override can be achieved as a backup to maintain confidence that the well will be controlled.

Currently Available Systems

Several manufacturers have developed automatic rig systems that are available to the drilling industry. The most advanced system is the Walker-Neer Apache rig that offers manual, remote-control, or fully automatic operation. By the early 1990s, 71 rigs had been produced and placed in operation over the past 15 years. A typical system is shown in Figure 9–13.

The Apache rig is designed to operate with two or three personnel and has been manufactured for operation up to approximately 6000 ft. Rigs operating up to 20,000 ft are available through an extension of its current design.The rig uses automatic pipe handling equipment to transfer single, 30–45-ft pipe joints to and from the rig floor, and uses automatic positioning, holding, and torquing devices for makeup and breakout of pipe connections during trips. Its documented record of accidents relating to pipe handling and rig floor activities has been *zero* over a five-year period.

Apache management has stated that moving loads for the rig are reduced by one-third due to its light weight and size and the unit offers a 50% reduction in rig-up/rig-down time over conventional rigs of similar size. Apache claims that spud-in can occur within four hours after equipment has been set in place at the rig site.

Apache indicates that connections of single pipe joints can be made in approximately one minute as compared to four minutes for conventional rigs. Complete roundtrips are comparable to conventional rigs at approximately

Fig. 9–13 High-speed top drive (Courtesy Parker Drilling Co.)

1000 ft/hr. The system is capable of handling small and large drillpipe of varying diameters as well as casing up to 26-in. diameter pipe. Apache also states that its routine drilling time is12%–15% less than that of conventional rigs and that its cost of operations is 8%–10% less.

The Varco Corporation has also made significant inroads into automating rig operations. It has elected to automate various segments of the rig, beginning with the top drive system.

Varco also offers automatic pipe-handling equipment as well as automatic torque wrenches used for making up and breaking out drill pipe as shown in Figure 9–14.

Parker Drilling Company through its Partech subsidiary has developed an automatic weight-on-bit control system ("Parkomatic") as well as ungeared top

Fig. 9–14 Automatic pipe wrenches (Courtesy Varco International, Inc.)

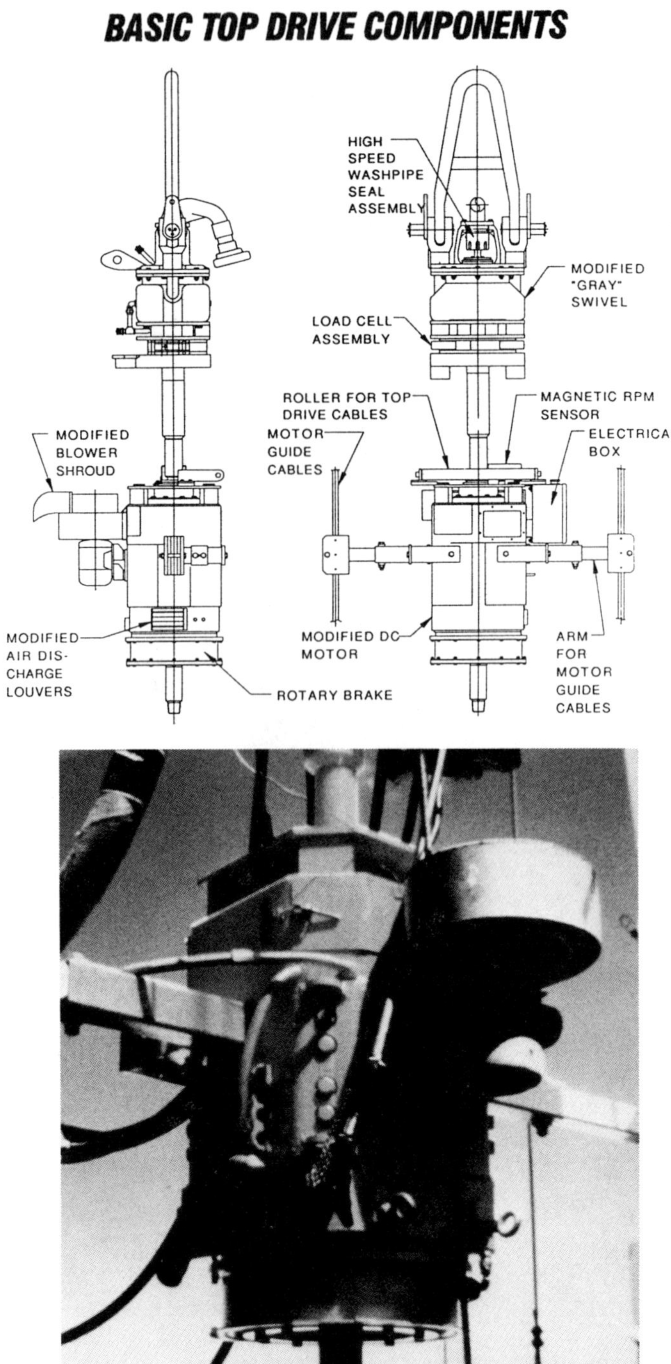

Fig. 9–15 Gearless direct top drive (Courtesy Parker Drilling Company)

Fig. 9–15 Gearless direct top drive (Courtesy Parker Drilling Company)

drives for coring operations for its drilling subsidiaries. The top drive system is shown in Figure 9–15.

Summary

The drilling industry is acknowledging that implementing automatic rig operations offers significant opportunities to improve safety, improve performance, and reduce hole loss. While some rig personnel would be removed from drilling operations, notably floor hands and pipe-handling personnel, other replacements would be required to handle the maintenance of the mechanical as well as the electrical systems. Therefore, rig automation should be applied on the basis of safety and performance, not for the purpose of eliminating personnel.

Several manufacturers are producing automated rig equipment from complete systems to individual functions for existing drilling rigs. Significant improvements in wellsite time, costs of operations, and improved drilling performance have led drilling contractors to purchase these systems. It has been acknowledged that producer/operators are using drilling contractors that can demonstrate these improvements with their systems.

Bibliography

Giannesini, Jean Francois. "Production technology for horizontal wells takes new direction", *World Oil,* May 1989.

Heagney, W. D., Burt G. Crawford. "Key to Safety: People, Not Automation." *Drilling Contractor,* December 1989/January 1990.

Parker Drilling Company, Tulsa, Oklahoma, Equipment Catalog on coring system, 1989.

Varco International, Inc. Houston, Texas, Equipment catalog, 1989.

Varco International, Inc. "Top Drive Drilling System Description and Specifications." Technical Bulletin, Orange, CA, 1988.

Wagner, Bob."Successful Hybridization of Drilling Technology For Reliable Continuous Coring Systems." Presented to Energy Technology Conference, Houston, Texas, 1988.

Wilson, J. E. and Frederick Bried. "Application of Programmable Logic Controllers for Pipeline Local and Remote Control", *IEEE Transactions On Industry Applications,* November/December, 1988.

W-N Apache Corporation, Wichita Falls, Texas, Catalog No. 188, Equipment description and specifications.

10 OPERATING EFFICIENTLY AND ECONOMICALLY

Most rig builders use electrical power for cost-saving benefits. Once the system is operational, the operator should outline a program to realize the efficiencies of the system and obtain feedback from operating personnel on the actual costs and performance they are achieving.

Power Load Utilization

As with any system that operates, energy costs are directly related to the rate at which energy is consumed. Fuel costs on drill rigs are directly related to the work being done; peripheral costs, such as maintenance and repair, are related to the operating time and stress placed on the equipment. In many cases, careful planning reduces costs by operating the equipment at the speeds and loads that are necessary to meet drilling needs. For example:

1. Operate mud pumps near full speed to optimize the power factor in the electrical system and eliminate the possibility of having to add an additional mud pump.
2. Use soft acceleration of the drawworks during tripping; little time is lost in the overall drilling program with slower acceleration rates but stress to equipment can be lowered.
3. Optimize rotary table speed and penetration rates to reduce power consumption.

4. Keep all unnecessary lighting, blowers, mixers and hotel loads off until needed.
5. Optimize the operating gear ratios of the drawworks to operate the maximum load conditions on the drive motor.

Engine-Generator Utilization

As noted earlier, diesel engines do not operate efficiently at power levels below 50% of their ratings. If the loads on the engine-generator sets are light, consider turning off a set if the remaining units will deliver the required kw and kva during a particular drilling interval. To improve maintenance costs, alternate each diesel engine set to even the wear rates on each one and prevent moisture accumulation.

Minimizing System Losses

Improperly maintained equipment develops friction and wear losses which are dissipated as heat and perform no useful work. Proper maintenance is essential to minimize operating and repair costs. Lubricating bearings, chains, and gears reduces system power losses and saves fuel.

Electrical losses are always directly proportional to the square of the current flowing in conductors, SCRs, transformers, motors, and other equipment. Adjusting motors and other electrical loads to minimize the current lessens system electrical losses and allows equipment to run cooler for longer life.

Increasingly, "energy-efficient" AC motors are being designed whose internal electrical losses are lower, thereby reducing the energy not producing useful work. An electric rig uses a variety of AC motors, and specifying energy-efficient motors reduces the auxiliary load losses on a rig.

Cost Accounting

The most effective technique available to rig managers who are concerned about cost control is an effective cost accounting system. During the rig proposal stage, a thorough economic analysis should be performed to justify the rig's construction. This analysis should be carried out as the initial basis for budgeting the rig's operation.

Knowing the cost of individual functions of a rig allows a historical record for tax and auditing purposes, a method of measuring performance of the operating crew, and a means of identifying excessive cost areas. Once cost-measuring techniques are implemented, continuous evaluation permits more cost-effective measures to improve operating profits.

The additional benefits a contractor can derive from an effective cost accounting system include accumulation of cost data to form a basis for future drilling or rig proposals and accurate equipment inventory to reduce theft losses and account for purchased equipment. They also include determining where price increases occur for labor, material, and equipment, as well as the amount of increases; and holding supervisory personnel accountable for their expenses and rewarding them when cost goals are met or exceeded.

The cost accounting technique establishes each rig as a profit center and places responsibility for its operation in the hands of a company manager. Budgets and goals are set annually, and periodic reviews (monthly or quarterly) are performed with company management. Analyzing fixed and variable operating costs plus the expenses incurred due to rig moves, idle periods, depreciation, and similar costs are essential in effective cost control and profit evaluation.

The basic elements used in a rig cost accounting system should include the following:

Chart of Accounts—All anticipated expenditures are coded with an appropriate numbering system. Each account number is used in paying invoices for labor, material, equipment, and other allocations.

Forecasting and Budgeting—Establishing an annual plan and budgeting for the plan allows the company to allocate its financial, equipment, and personnel resources. Adjustment of the plan and budget is normally required as the year unfolds and variances occur.

Cost Retrieval System—Methods for obtaining accurate cost information should be established and proper account assignments made.

Cost Accounting Policy—This establishes the method of depreciation for such cost items as G&A allocations, insurance allocation, idle rig costs, and similar costs not under the direct control of the rig manager.

Responsibility and Authority—Management must be assigned responsibility for rig operations with appropriate authority extended.

The International Association of Drilling Contractors (IADC) has established its accounting manual for rig contractors and includes formats and a chart of accounts that are quite helpful.

A cost-accounting system requires considerable discipline; policies and procedures must be enforced to be effective. Constant review of costs along with corrective action allows the drilling contractor to improve profits with an electrical system and create data for the next rig or drilling proposal.

Bibliography

Parsons, Richard C. "Local Cost Control Improves Rig Profits." *World Oil*, June 1979.

11 ELECTRIC EQUIPMENT MAINTENANCE

The effectiveness of a rig maintenance program influences operating costs, repair costs, downtime, and replacement costs. Regularly scheduled equipment maintenance optimizes profits and maintains a drilling record with minimum downtime.

In some areas of rig maintenance, further guidance normally is supplied by equipment manufacturers. System reliability is directly related to the condition of the equipment and the service it receives. Irreparable damage can occur to electrical equipment if operated when there is an indication of trouble or when dust accumulation, for example, is severe.

Maintaining a single piece of a complete system's equipment at the expense of others is also a poor policy since all equipment must function well together. Neglect must be avoided and regular maintenance procedures established and followed.

Predictive Maintenance vs Preventive Maintenance

A common phrase used throughout industry is the term "preventive maintenance," which means a regularly scheduled program of inspections and repair. Also used is "predictive maintenance," which means to repair and replace based on historical wear information. Some equipment is more adaptable to one form of maintenance than another, and an evaluation of the equipment can determine which program is best. On drilling rigs, both programs can be combined to optimize the maintenance record of equipment.

To understand these two philosophies, consider a medium-sized electric land drilling rig. Its equipment list includes DC drilling motors, diesel engine-AC generator sets, SCR controls, and generator controls. To further analyze the equipment for maintenance, list each function and separate those items in the equipment list that follow this criteria:

Preventive Maintenance Items—Life and wear history are unknown or uncertain and are based upon the care, cleanliness, coolant, lubrication, and other factors under control of the maintenance personnel.

Predictive Maintenance Items—Life and wear history are known and depend primarily upon number of cycles, operations, or time.

Based on these definitions, the land rig system would be maintained as in this example:

DC motors

- Preventive maintenance items
  - Lubricate bearings monthly
  - Clean blower vents weekly
  - Check mounting bolts weekly
  - Tighten electrical connections weekly
- Predictive maintenance items
  - Brushes have 10,000-hour life; replace every 9000 hours
  - Bearings have a B-10 life of 18,000 hours; replace every 15,000 hrs.
  - Overhaul motor every three years under heavy-duty operation; overhaul unit every four years under light-duty operation

SCR controls

- Preventive maintenance items
  - Clean dust from equipment daily
  - Check cooling blower operation weekly
  - Check all instruments daily
  - Check all load assignments weekly
- Predictive maintenance items
  - DC contactors have a mechanical life of 10,000 operations and cycle on-off 50 times a day on the average. Replace the contacts after (10,000/50) or 200 days of operation during regularly scheduled downtime.

A particular circuit breaker has a mechanical life of 1000 operations and is operated at least three times a week; replace after (1000/3 x 52) or six years.

A similar analysis can be made on all rig equipment, including such mechanical items as drill bits, drawworks, rotary table, and traveling block. The importance of a predictive along with a preventive maintenance program is the overall coverage and care of rig equipment. Preventing an item from failing while the system is operating saves repair and downtime at an inopportune moment and also prevents consequential damage to other equipment caused by this failure.

Embarking on such a program requires equipment knowledge, advance planning, and discipline to maintain the schedule. However, the rewards in terms of reduced operating and repair costs are significant over the long term.

Diesel Engines

The American Petroleum Institute has produced a recommended practice procedure (API RP 7C-11F) titled "Installation, Maintenance and Operation of Internal Combustion Engines" from which the following excerpt is taken. (The maintenance supervisor must define those areas that should be classified preventive maintenance and plan accordingly.)

Recommended Maintenance Practice (Courtesy, API)

Internal Combustion Engines

24. Daily Inspection. A daily inspection of engines should be made by the pumper or qualified engine operator who should also record each inspection performed as follows:

a. Coolant temperature and oil pressure. Before shutting down, coolant temperature and oil pressure should be observed under normal load. If the engine is a diesel equipped with pyrometers, all cylinders should be checked. Check and record vacuum reading on spark ignition engines.

b. Lubrication. The crankcase oil level in both main and starting engine should be checked and oil added if needed. Sufficient time should be allowed for the oil to drain back to the crankcase from the upper engine parts before checking the oil level. Oil should be replaced as recommended by the engine manufacturer or when it is obviously thickened, diluted, or dirty beyond normal conditions. Filter elements should be replaced and the oil strainer cleaned during the change of oil or more frequently if recommended by the engine manufacturer. All points recommended by the manufacturer as requiring daily attention should be checked, but care should be exercised that the proper lubricants are used and that these points are not overlubricated.

c. *Cooling system.* Coolant-water supply should be checked, and clean, soft water added as needed. In water, lost coolant should be replaced with the proper anti-freeze mixture. The cooling system vent valve, when provided, should be opened before adding water and closed after water has been added. Water showing rust or scum may indicate a need for replacement of the cleaning system, hose, or gaskets. A manufacturer-approved cooling water treatment agent should be used. On air-cooled engines, the flywheel air screen and air intake stack should be checked, and any foreign material removed. If flywheel air screen or intake stack is very dirty, the fins on heads and cylinder blocks should be inspected and cleaned if necessary. If cylinder block fins are rusty, they should be cleaned with a wire brush.

d. *Air cleaners.* Air cleaners on engine air-supply and breather openings should be checked and cleaned as required according to the design and condition of the cleaner. Pay particular attention on turbo- charged engines to any indications of metallic filtering media breaking up or holes and breaks in paper element cleaners.

e. *Fuel supply system.* The fuel supply system should be checked by draining the sump traps and strainers; also, the fuel-oil line filters, if the engine is a diesel. All collections of dirt and dust around the tank filler caps and vents should be removed.

f. *Leaks or damage.* A visual inspection should be made of all water, fuel, lubricant lines, fittings, and valves for indications of leaks or damage. Special attention should be directed to turbo-charger oil and water lines, mountings, and air and exhaust line connections and seals.

g. Daily engine report. All work done, the hours of engine operation, and the amounts of oil, antifreeze, and special lubricants should be recorded on the daily engine report.

h. Malfunctioning or needed repair. Any malfunctioning of the engine or any needed repairs should be reported to the mechanic.

25. Weekly Inspection. The following weekly inspection of engines should be made by the pumper or qualified engine operator, who should also record each inspection performed.

a. Daily inspection items. All of the daily inspection items as given in Par. 24 should also be performed in the weekly inspection.

b. Cleaning. If necessary, the engine exterior should be thoroughly cleaned with a mild solvent (not gasoline), and compressed air or hot water should be used for flushing and drying. Care should be taken not to wash or blow dirt into inaccessible locations behind filler openings or into ignition or injection equipment; or on air-cooled engines, into the fins on heads, and cylinder blocks. Always dry and relubricate governor and control linkage joints after flushing engine.

c. Water pump. Water pump seals should be inspected and packing on packed-type pumps should be tightened or replaced if necessary.

d. Air Cleaners. Air cleaners on engine air-supply and breather openings should be checked and cleaned as required according to the design and condition of the cleaner. Pay particular attention on turbo-charged engines to any indications of metallic filtering media breaking up or holes and breaks in paper element cleaners.

e. Fuel-Supply System. The fuel-supply system should be checked by draining the sump traps and strainers; also the fuel-oil line filters, if the engine is a diesel. All collections of dirt and dust around thetank filler caps and vents should be removed.

f. Leaks or Damage. A visual inspection should be made of all water, fuel, lubricant lines, fittings, and valves for indications of leaks or damage. Special attention should be directed to turbo-charger oil and water lines, mountings, and air and exhaust line connections and seals.

g. *Daily Engine Report.* All work done, the hours of engine operation, and the amounts of oil, antifreeze, and special lubricants used should be recorded on the daily engine report.

h. *Malfunctioning or Needed Repair.* Any malfunctioning of the engine or any needed repairs should be reported to the mechanic.

25. Weekly Inspection. The following weekly inspection of engines should be made by the pumper or qualified engine operator who should also record each inspection performed.

a. *Daily Inspection Items.* All of the daily inspection items as given in Par. 24 should also be performed in the weekly inspection.

b. *Cleaning.* If necessary, the engine exterior should be thoroughly cleaned with a mild solvent (not gasoline), and compressed air or hot water should be used for flushing and drying. Care should be taken not to wash or blow dirt into inaccessible locations behindfiller openings or into ignition or injection equipment; or on air-cooled engines, into the fins on heads, and cylinder blocks. Always dry and relubricate governor and control linkage joints after flushing engine.

c. *Water Pump.* Water pump seals should be inspected, and packing on packed-type pumps should be tightened or replaced if necessary.

d. *Fan Belts.* Fan belts should be checked for proper tension and tightened or loosened if needed.

e. *Lubrication of Generator and Accessories.* The generator should be lubricated with a few drops of oil and the accessories lubricated according to the instructions of the manufacturer.

f. *Power Take-Off Clutch.* The power take-off clutch should be lubricated and adjusted, if required, according to the instructions of the manufacturer; however, care should be taken not to overlubricate.

g. *Gas Regulators.* Gas engines should be checked for gas pressure at the primary and final regulators.

h. *Breather Elements.* All removable breather elements should be carefully cleaned and washed in mild solvent (not gasoline) and the oil changed on those elements requiring re-oiling. Follow instructions carefully on dry type element service.

i. Spark Plugs. Spark plugs should be removed, inspected, and regapped, if necessary, to the recommended setting.

j. Fuel Strainers. Diesel fuel system strainers should be cleaned and filters replaced as scheduled.

26. Monthly Inspection. The following monthly inspection should be performed by an expert mechanic who should also record each inspection performed.

a. Daily and Weekly Inspection Items. All the inspection items given in Par. 24 and 25 should also be performed in the monthly inspection.

b. Ignition System. On spark-ignition engines, the following ignition devices, depending on the type used, should be checked:

(1) magnetopoint condition and clearances timing

(2) the impulse functioning

(3) the sparkplug gap and heat range

(4) the distributor condition with respect to the automatic advance mechanism

c. Valves. The external appearance of the valve mechanism should be checked, as well as the condition of valve springs and the proper lubrication of valve rockers, push-rod ends, and valve stems. All valve clearances should be set according to the instructions of the engine manufacturer. Valve timing should be checked if an adjustable timing device is provided. The compression on all cylinders should be measured if the engine lacks power or if the condition of the valves and rings is questionable. The functioning of the compression-release device should be checked on diesel engines if it is used. Engines using hydraulic valve lifters should be checked for sounds of lifter malfunction and the manufacturer's inspection procedure followed.

d. Starting Equipment. The starting equipment should be carefully tested and inspected. Starting engines should be checked for lubrication and general condition, with special attention given to the mounting bolts, bendix-drive lubrication, engagement linkage, pinion gear teeth mesh and adjustment, and fuel tank strainer.

Manufacturer recommendations for specific makes and types of engines should be observed. Add the recommended lubricant to air starter lubricant reservoirs and clean air traps of dirt.

e. *Engine Mounts.* Engine mounts should be inspected and tightened if required. A check should be made for signs of engine shifting, misalignment, loosening of coupling or pulley, or improper loading. Any shifting should be corrected and all points of alignment rechecked.

f. *Cooling Fan.* The cooling fan should be examined for evidence of physical damage or cracking in the hub or spider area. If the fan hub bearings require lubrication by disassembly and packing or by installation of a special grease fitting, this operation should be performed.

g. *Safety Shields.* All fan belt and shaft safety shields should be repaired and reinstalled.

h. *Rocker Covers and Inspection Doors.* New gaskets should be used on all rocker covers and inspection doors.

i. *Seasonal Check of Cooling System.* Particularly at the changes of season and when starting to use or remove antifreeze, the cooling system should be flushed thoroughly. The thermostats should also be removed and tested for correct functioning. Evidence of scale, sludge, or rust deposits in the cooling system warrants further investigation, and a special cleaning of oil coolers and heat exchangers may be necessary.

j. *Crankcase.* Inspection plates should be removed, if the crankcase is so equipped, and a check made for sludge in the crankcase. The oil-pump screen should be checked and cleaned if necessary.

k. *Safety Devices, Generator, and Battery.* A check should be made of safety devices, generator, and battery (or gas starter, if used). Check the actual function of overtemperature, low oil pressure, and overspeed shutdowns.

l. Vibration Damper. Inspect the vibration damper for damage, runout, signs of deterioration or loss of viscous material, or looseness.

m. Turbocharger Impeller. Inspect turbocharger compressor impeller for accumulations of dirt, dust, and oil. Clean according to manufacturer's recommendations.

n. Records. The work done, materials used, and time required should be recorded.

Diesel Engines

44. Starting Difficulty. If a diesel engine fails to start or does not start readily, the following possible causes of trouble should be checked in an effort to locate the difficulty.

a. Fuel failure, low-pressure side

(1) Line valves not open; tank empty

(2) Ice in lines or traps

(3) Cold fuel

(4) Plugged fuel filters, or dirt in lines between filter and pump

(5) Fuel tank too low in relation to transfer pump

(6) Dirt under transfer-pump valves or worn valves

(7) Air lock in fuel pump or injection pump

(8) Safety switch not being held open by operator

Note: The foregoing items may be checked by opening the bleeder valve and cranking the engine. A pressure gauge should be used in the bleeder- valve hole to check primary pump pressure. A hand plunger may be used on the transfer pump, if desired. A substantial flow of fuel without air bubbles should exit from the bleeder opening.

b. Fuel failure, high-pressure side

(1) Enrichment lever not in proper position; rack partly closed in cold weather

(2) Stop control in wrong position

(3) Air locks in high-pressure lines

(4) Broken or disconnected pump-drive coupling

Note: The foregoing items may be checked by loosening line-coupling nuts a few turns at each nozzle and cranking engine. A substantial flow of fuel should occur at each injection impulse. If no fuel appears and an adequate flow of fuel is known to have reached the plungers, either the plungers or delivery valves may be stuck as a result of poor fuel, improper storage, or inadequate lubrication.

c. *Poor nozzle spray pattern or gummed or corroded nozzles*

d. *Faulty injection timing*

e. *Glow plugs too cold*

f. *Battery voltage low* (A fully charged 12-volt heavy-duty battery at normal temperatures will show 10.5 volts while cranking.)

g. *Air cleaners clogged or protective covers accidentally drawn into the manifold*

h. *Poor compression* (check each cylinder)

i. *Liquid lock between piston crown and cylinder head due to flushing oil from storage, leaking head gasket, or leaking injector*

j. *Low cranking speed* due to weak batteries, poor starter condition, or thick, cold oil

45. Engine Stops Running. If the diesel engine suddenly stops running, the following possible causes of trouble should be checked in an effort to locate the difficulty.

a. *Lack of fuel*

b. *Fuel lines obstructed or broken*

c. *Automatic low oil-pressure or high water-temperature safety control may have operated*

d. *Excessive overload or improper governor adjustment*

e. *Plugged fuel-tank vent*

f. *Damaged transfer or injection pump drive*

46. Low Power. If the diesel engine has low power and runs unevenly, the following possible causes of trouble should be checked in an effort to locate the difficulty.

a. Inadequate supply of fuel to pump
b. Fuel-tank vent partially plugged
c. Faulty timing
d. Delivery valves not operating properly
e. Dirty or damaged injection plunger
f. Leaking fuel lines or air in lines
g. Damaged or excessive clearance in blowers
h. Overflow valve or injector drain line feeding back into primary pump inlet
i. Dirty or clogged nozzles
j. Air cleaner or manifold obstructed
k. Row or uneven compression
(1) Broken valve spring
(2) Sticking valves
(3) Badly worn rocker arms
(4) Sticking cam followers
(5) Bent throttle control linkage
(6) Binding of injector-rack control tube or injector racks
l. Fuel oil not to specification
m. Improper exhaust line
n. Leaking turbocharger air connections
o. Dirty or damaged turbocharger
p. Improper intercooler operation

47. Surging or Irregular Speed. If the diesel engine develops a surging or irregular speed, the following possible causes of trouble should be checked in an effort to locate the difficulty.

a. Governor needs lubrication
b. Improper grade of governor oil (hydraulic governors)
c. Governor improperly adjusted
d. Injection pump

(1) Lack of lubrication

(2) Insufficient fuel supply for primary system

(3) Irregular operation of automatic bleeder valve; air entrainment in pump and lines, valves, or nozzles

(4) Inaccurate pump timing

e. *Slipping clutch or belt drive; wide variation in loads or poor regulation on electrical equipment*

f. *Dirty or damaged turbocharger system*

48. Overheating. If the diesel engine overheats, the following possible causes of trouble should be checked in an effort to locate the difficulty.

a. *Excessive exhaust back pressure*

(1) Restricted muffler or loose baffles in muffler

b. *Cooling system*

(1) Insufficient coolant

(2) Radiator frozen or clogged tubes and tanks

(3) Radiator core dirty (external)

(4) Water hose clogged

(5) Slipping fan belt

(6) Thermostat stuck

(7) Cooling system inadequate

(8) Improper air recirculation

(9) Aeration of water from leaking gaskets or pump

(10) Defective water pump

(11) Excessive back pressure on external cooling system

(12) Air shroud, air stack, cylinder-head fins, or cylinder-block fins blocked with debris

c. *Combustion*

(1) Improper fuel

(2) Faulty injection timing, retarded or wrong cycle

(3) Faulty injection nozzles

(4) Pump setting incorrect

d. Lubrication

(1) Improper or excessive time between oil changes

(2) Air-locked or plugged oil filter, cooler, or screen

e. Load

(1) Prolonged service at excessive load

(2) Improper synchronization of two or more engines

f. Installation

(1) High-exhaust back pressure due to improper piping or muffling

(2) Insufficient air circulation when engines are operating in closed spaces

(3) Improper turbocharging; intercooler too hot

49. Low or Fluctuating Oil Pressure. If the diesel engine develops a low or fluctuating oil pressure, the engine should be stopped at once and the following possible causes of trouble should be checked in an effort to locate the difficulty.

a. Oil

(1) Insufficient oil

(2) Dirty filters, oil coolers, or sump screen

(3) Improper grade of oil

(4) Foaming oil due to water leakage

b. Pressure regulation

(1) Worn, sticking, or loose relief valve

(2) Vent behind relief valve plugged

(3) Inaccurate pressure gauge

c. Pump

(1) Inlet-strainer screen clogged

(2) Damaged or worn pump gears

(3) Clogged oil lines or passages

d. Mechanical

(1) Excessive bearing clearances on camshaft or crankshaft

(2) Cracked or broken oil line

(3) Leaking gasket on suction side

50. Knocking or Unusual Noises. If the diesel engine starts knocking or develops other unusual noises, the following possible causes of trouble should be checked in an effort to locate the difficulty.

a. Operation

(1) Unsatisfactory fuels

(2) Overloading (heavy black smoke)

(3) Improper timing

b. Installation

(1) Engine loose on mounts

(2) Vibrational disturbances of loose control rods, air cleaner, muffler, or similar parts

(3) Loose accessories such as generators, fans, compressors, etc.

c. Mechanical

(1) Loose bearings on connecting rods, piston pins, camshaft, or crankshaft

(2) Loose flywheel or distorted housing

(3) Damage, looseness, or wear in water-pump, oil-pump, or fuel-pump drive

(4) Excessive crankshaft end play

(5) Improperly adjusted valves, sticking valves, rocker arms, or tappets

(6) Excessive time since overhaul (worn pistons, stuck or broken rings, carbon on piston crown)

(7) Defective vibration damper

AC Machines

AC motors can be synchronous, induction, or wound rotor; polyphase or single phase; and fractional or integral horsepower sizes. AC synchronous generators use either brush or brushless field exciters.

All electrical power should be removed before maintenance on any machine is performed.

Preventive Maintenance

Monthly:

1. Check for mechanical stability under operation.
2. Clean motor interior with dry, clean, compressed air and remove any oil or grease.
3. Inspect slip rings or exciter commutator for wear. Clean, adjust, and resurface as required.
4. Examine armature and field coils and lead connections for faults, excessive heat, charring, or cracks.
5. Lubricate bearings according to manufacturer's instructions.
6. Annually inspect the armature and field coil insulation resistance with a megohmmeter according to manufacturer's instructions.

Predictive Maintenance

(Obtain component life data from manufacturer.)

1. Replace bearings.
2. Replace brushes, holders, and slip rings.
3. Replace commutator and mica insulators.
4. Overhaul machine.
5. Revarnish and bake winding.
6. Repack lubricated bearings.
7. Replace blower and motor.

Tables 11–1 and 11–2 give a checklist for isolating faults in AC motors and AC generators.

DC Machines

DC machines contain several components which require regular maintenance. Along with the maintenance program, the machine must be operated within its rating to prevent overheating and stress. This way, excessive speeds and accelerating currents are avoided. Always remove electrical power before performing maintenance. Troubleshooting aids are shown in Table 11–3.

Preventive Maintenance

Monthly:

1. Clean motor interior with dry, clean, compressed air and remove any oil or grease.
2. Inspect commutator for a smooth, polished appearance. Clean and resurface as required.

Table 11–1
AC Motor Troubleshooting Chart

Trouble	Cause	What To Do
Motor fails at start	Blown fuses	Replace fuses at least 125% of nameplate amperes.
	Overload trips	Check and reset overload in starter.
	Improper current	Check to see that power supply supplied agrees with motor nameplate and load factor.
	Improper line connections	Check connections with diagram supplied with motor.
	Open circuit in winding or starting switch	Indicated by humming sound when switch is closed. Repulsion induction motors may spark at brushes. Check for loose wiring connections; also see if starting switch inside motor is closed.
	Mechanical failure	Check to see if motor and drive turn freely. Check bearings and lubrication.
	Short-circuited stator	Indicated by blown fuses. Rewind motor.
	Poor stator coil connection	Remove end bells; locate with test lamp.

Trouble	Cause	What To Do
	Rotor defective	Look for broken bars or end rings.
	Motor may be overloaded	Reduce load.
	If three phase, one phase may be open	Check lines for open phase.
	Defective capacitor	Check for short circuit, grounded or open capacitor; replace if necessary.
	Worn or sticking brushes on repulsion induction motors	Check for wear and correct brush pressure. Clean commutator if dirty.
Motor stalls	Wrong application	Change type or size. Consult manufacturer.
	Overloaded motor	Reduce load.
	Low motor voltage	See that nameplate voltage is maintained.
	Open circuit	Fuses blown, check overload relay, stator, and pushbuttons.
	Incorrect control resistance of wound rotor	Check control sequence. Replace broken resistors. Repair open circuits.
Motor runs and then dies down.	Power failure	Check for loose connections to line, to fuses, and to control.
Motor over-heats while running under load	Low voltage	Check terminals of motor with a voltmeter.
bearings	Rotor rubs stator bore	If not poor machining, replace worn bearings.

Table 11–2
AC Generator Troubleshooting Chart

Trouble	Cause	What to Do
Generator fails to develop proper voltage	Blown fuses	Replace fuses on output, if used, at least 125% of name-plate amperes.
	Improper output	Check to see that generator rating agrees with load and power factor.
	Improper phase sequence	Check connections with diagram supplied with generator; reverse any two-phase connections to correct.
	Open circuit in winding	Check for loose wiring connection; also see if windings have continuity.
	Mechanical failure	Check to see if generator turns freely. Check bearings and lubrication.
	Short-circuited stator	Indicated by overheating on one phase.
	Poor connection	Remove junction box cover and inspect.
	Rotor defective	Look for defective field winding.
Generator vibrates after corrections have been made	Generator mis-aligned	Realign.
	Weak foundations	Strengthen base.
	Coupling out of balance	Balance coupling.
	Driving equipment unbalanced	Rebalance driving equipment.
	Defective ball bearing	Replace bearing.
	Bearings not in line	Line up properly.

Trouble	Cause	What to Do
	Balancing weights shifted	Rebalance rotor.
	Polyphase generator running single phase	Check for open circuit.
	Excessive end play	Adjust bearing or add washer.
Unbalanced line current on polyphase generators during normal operation	Unequal terminal volts	Check leads and connections.
	Single-phase operation	Check for open contacts.
Scraping noise	Loose on bedplate	Tighten holding bolts.
Noisy operation	Airgap not uniform	Check and correct bracket fits or bearing.
	Rotor unbalance	Rebalance.
	Deterioration of grease or lubricant contaminated	Remove old grease, wash bearings thoroughly in kerosene, and replace with new grease.
	Excess lubricant	Reduce quantity of grease; bearing should not be more than one-half filled.
	Overloaded bearing	Check alignment, side, and end thrust.
	Broken ball or rough races	Replace bearing; first clean housing thoroughly.
Motor does not come up to speed	Not applied properly	Consult supplier for proper type.
	Voltage too low at motor terminals because of line drop	Use higher voltage on transformer terminals or reduce load.
	If wound rotor, improper control operation of secondary resistance	Correct secondary control.

Trouble	Cause	What to Do
	Starting load too high	Check load motor will carry at start.
	Broken rotor bars	Look for cracks near rings. A new rotor may be required as repairs are usually temporary.
	Open primary circuit	Locate fault with testing device and repair.
Motor takes too long to accelerate	Excess loading	Reduce load.
	Poor circuit	Check for high rotor resistance.
	Defective squirrel cage rotor	Replace with new rotor.
	Applied voltage too low	Check transformer output.
Wrong rotation	Wrong sequence of phases	Reverse connections at motor or at switchboard.
Motor over-heats while running under load	Overload	Reduce load.
	Wrong blowers or air shields may be clogged with dirt and prevent proper ventilation of motor	Good ventilation is manifest when a continuous stream of air leaves the motor. If it does not after cleaning, check manufacturer.
	Motor may have one phase open	Check to make sure that all leads are well connected.
	Grounded coil	Locate and repair.
	Unbalanced terminal voltage	Check for faulty leads, connections, and transformers.
	Shorted stator coil	Repair and check wattmeter reading.
	Faulty connection	Indicated by high resistance

Trouble	Cause	What to Do
	High voltage	Check terminals of motor with a voltmeter.
Motor vibrates after corrections have been made	Motor misaligned	Realign.
	Weak foundations	Strengthen base.
	Coupling out of balance	Balance coupling.
	Driven equipment unbalanced	Rebalance driven equipment.
	Defective ball bearing	Replace bearing.
	Bearings not in line	Line up properly.
	Balancing weights shifted	Rebalance rotor.
	Wound rotor coils replaced	Rebalance rotor.
	Polyphase motor running single phase	Check for open circuit.
	Excessive end play	Adjust bearing or add washer.
Unbalanced line current on polyphase motors during normal operation	Unequal terminal volts	Check leads and connections.
	Single-phase operation	Check for open contacts.
	Poor rotor contacts in control wound rotor resistance	Check control devices.
	Brushes not in proper position on wound rotor	See that brushes are properly seated and shunts in good condition.
Scraping noise	Fan rubbing air shield	Remove interference.
	Fan striking insulation	Clear fan.
	Loose on bedplate	Tighten holding bolts.

Trouble	Cause	What to Do
Noisy operation	Airgap not uniform	Check and correct bracket fits or bearing.
	Rotor unbalance	Rebalance.
Hot bearings general	Bent or sprung shaft	Straighten or replace shaft.
	Excessive belt pull	Decrease belt tension.
	Pulleys too far away	Move pulley closer to motor bearing.
	Pulley diameter too small	Use larger pulleys.
	Misalignment	Correct by realignment of drive.
Hot bearings sleeve	Oil grooving in bearing obstructed by dirt	Remove bracket or pedestal with bearing and clean bearing housing and oil grooves.
	Oil too heavy	Use recommended lighter oil.
	Oil too light	Use recommended heavier oil.
	Too much end thrust	Reduce thrust induced by drive, or supply external means to carry thrust.
	Badly worn bearing	Replace bearing.
Hot bearings ball	Insufficient grease	Maintain proper quantity of grease in bearing.
Hot bearings general	Bent or sprung shaft	Straighten or replace shaft.
	Excessive belt pull	Decrease belt tension.
	Pulleys too far away	Move pulley closer to generator bearing.
	Pulley diameter too small	Use larger pulleys

Trouble	Cause	What to Do
	Misalignment	Correct by realignment of drive.
Hot bearings sleeve	Oil grooving in bearing obstructed by dirt	Remove bracket or pedestal with bearing and clean bearing housing and oil grooves.
	Oil too heavy	Use recommended lighter oil.
	Oil too light	Use recommended heavier oil.
	Too much end thrust	Reduce thrust induced by drive, or supply external means to carry thrust.
	Badly worn bearing	Replace bearing.
Hot bearings ball	Insufficient grease	Maintain proper quantity of grease in bearing.
	Deterioration of grease or lubricant contaminated	Remove old grease, wash bearings thoroughly in kerosene, and replace with new grease.
	Excess lubricant	Reduce quantity of grease; bearing should not be more than one-half filled.
	Overloaded bearing	Check alignment, side, and end thrust.
	Broken ball or rough races	Replace bearings; first clean housing thoroughly.
Generator overheats while running under load	Overload	Reduce load.
	Wrong blowers or air shields may be clogged with dirt and prevent proper ventilation	Good ventilation is manifest when a continuous stream of air leaves the generator. If it does not after cleaning, check manufacturer.

Trouble	Cause	What to Do
	Grounded coil	Locate and repair.
	Unbalanced terminal voltage	Check for faulty leads and connections.

Table 11–3
DC Motor Troubleshooting Chart

Trouble	Cause	What to Do
Motor fails to start	Blown fuses	Replace fuses in control equipment.
	Overload trips	Check and reset overload in controller.
	Improper current supply	Check to see that power supplied agrees with motor nameplate rating.
	Improper line connections	Check connections with diagram supplied with motor.
	Open circuit in winding	Check for loose wiring connections.
	Mechanical failure	Check to see if motor and drive turn freely. Check bearings and lubrication.
	Short-circuited armature	Indicated by blown fuses. Motor must be rewound.
	Poor connection	Remove junction box cover and locate.
	Rotor defective	Look for broken commutator bars or brushes.
	Motor may be overloaded	Reduce load.
	Worn or sticking brushes	Check for wear and correct brush pressure. Clean commutator if dirty.

Trouble	Cause	What to Do
Motor stalls	Wrong application	Change type or size. Consult manufacturer.
	Overloaded motor	Reduce load.
	Low motor voltage	See that correct voltage is maintained.
	Open circuit	Fuses blown, check controller.
Motor runs and then dies down	Power failure	Check for loose connections to line, to fuses, and to controller.
Motor overheats while running under load	Overload	Check current to armature
	Rotor rubs stator bore	If not poor machining, replace worn bearings.
Motor vibrates after corrections have been made	Motor misaligned	Realign.
	Weak foundations	Strengthen base.
	Coupling out of balance	Balance coupling.
	Driven equipment unbalanced	Rebalance driven equipment.
	Defective ball bearing	Replace bearing.
	Bearings not in line	Line up properly.
	Balancing weights shifted	Rebalance rotor
	Wound rotor coils replaced	Rebalance rotor
	Polyphase motor running single phase	Check for open circuit.
	Excessive end play	Adjust bearing or add washer.

Trouble	Cause	What to Do
Noisy operation	Airgap not uniform	Check and correct bracket fits or bearing.
	Rotor unbalance	Rebalance.
Hot bearings general	Bent or sprung shaft	Straighten or replace shaft.
	Excessive belt pull	Decrease belt tension
	Pulleys too far away	Move pulley closer to motor bearing.
	Pulley diameter too small	Use larger pulley.
	Misalignment	Correct by realigning drive.
Hot bearings sleeve	Oil grooving in bearing obstructed by dirt	Remove bracket or pedestal with bearing and clean bearing housing and oil grooves; renew oil.
	Oil too heavy	Use recommended lighter oil.
	Oil too light	Use recommended heavier oil.
	Too much end thrust	Reduce thrust induced by drive, or supply external means to carry thrust.
	Badly worn bearing	Replace bearing.
Hot bearings ball	Insufficient grease	Maintain proper quantity of grease in ball bearing.
	Deterioration of grease or lubricant contaminated	Remove old grease. Wash bearings thoroughly in kerosene, and replace with new grease.
	Excess lubricant	Reduce quantity of grease; bearing should not be more than one-half filled.

Trouble	Cause	What to Do
	Overloaded bearing	Check alignment, side, and end thrust.
	Broken ball or rough races	Replace bearing; first clean housing thoroughly.
Motor does not come up to speed	Not applied properly	Consult supplier for proper type.
	Voltage too low at motor terminals because of line drop.	Use higher voltage or reduce load.
	Starting load too high	Check load motor will carry at start.
	Open field, if shunt	Locate fault with ohmmeter and repair.
Motor takes too long to accelerate	Excess loading	Reduce load.
	Poor circuit	Check for high resistance.
Wrong rotation	Wrong field polarity	Reverse connections at motor or at switchboard.
Motor overheats while running under load	Overload	Reduce load.
	Wrong blowers or shields, may be clogged with dirt, preventing proper ventilation	Good ventilation is manifest when a continuous stream of air leaves the motor. If it does not after motor cleaning, check manufacturer.
	Grounded coil	Locate and repair.
	Faulty connection	Indicated by high resistance

3. Check brush holder clips and mechanism, connecting wire, and insulators. Replace as required.
4. Check brushes for wear, broken or chipped surfaces, and length. Replace as required and check for free movement in brush holders.

5. Check and adjust brush spring pressure following manufacturer's suggested practice.
6. Examine armature and field coils and lead connections for faults, excessive heat, charring, or cracks.
7. Lubricate bearings according to manufacturer's recommendation.
8. Annually inspect the armature and field coil insulation resistance with a megohmmeter following manufacturer's instructions.
9. Check for mechanical stability under operation.

Predictive Maintenance

(Obtain component life data from manufacturer.)

1. Replace bearings.
2. Replace brushes and holders.
3. Replace commutator and mica insulators.
4. Overhaul motor.
5. Revarnish and bake winding.
6. Repack lubricated bearings.
7. Replace blower and motor.

Electrical Switchgear

Switchgear equipment includes items containing circuit breakers, starters, relays, contactors, switches, pushbuttons, and other electromechanical devices. This equipment controls power flow to motors, transformers, and other electrical gear on an electric drill rig or vessel.

From an operations viewpoint, those items that wear from repeated use, current arcing, or heat are prime candidates for periodic maintenance. Keeping equipment clean and dry is the first maintenance requirement, followed by component replacement as wear occurs. All power should be removed from equipment during maintenance.

Preventive Maintenance

1. Keep all switchgear and components clean and free from dust, particles, and other foreign matter with vacuum cleaner.

2. Inspect all current-carrying contacts for wear and pitting.
3. Inspect all return springs and mechanisms on moving devices.
4. Check all panel-mounted components such as meters, switches, and lamps and replace as needed.
5. Inspect for proper equipment grounds.
6. Inspect and megger-test power cables to switchgear.
7. Inspect bus and support insulators.
8. Inspect all metering and control transformers for discoloration or damage.
9. Check condition of wiring and terminations.
10. Correct unsafe conditions.
11. Remove and check drawout circuit breakers.

Predictive Maintenance

1. List all contactors, starters, and other power-handling devices; determine the mechanical and electrical life of their contacts from vendors. Determine the projected life and replace contacts prior to anticipated failure. Experience will define this replacement period as the equipment is used.
2. List all circuit breakers and similar devices; determine the mechanical and electrical life of the contacts and operating mechanisms from vendors. Replace these items prior to projected failure point.
3. Replace all lamps and indicator bulbs periodically, based on projected lamp life.
4. Evaluate other components with limited life and replace prior to predicted failure.

SCR Equipment

The most significant advantage of solid-state SCR equipment is the lack of moving parts and resulting low maintenance required. Only such items as panel-mounted components, bus bars, fans, and connectors require maintenance. General cleanliness assures good heat transfer of the SCR heat sinks and reduces the possibility of arcing between live bus bars and components. All power should be removed from equipment during maintenance.

Preventive Maintenance

1. Vacuum and clean equipment regularly to prevent dust and oil film accumulation on components.
2. Inspect all bus bars and SCR heat sinks for proper tightness at connection points.
3. Inspect for proper compression of heat sinks on SCRs.
4. Check all panel-mounted components and replace as needed.
5. Inspect for proper equipment grounds.
6. Inspect all metering and control transformers for discoloration or damage.
7. Check condition of wiring and terminations.
8. Inspect cooling fan for proper air flow.

Predictive Maintenance

1. Replace all lamp bulbs prior to predicted failure.
2. Replace cooling fan prior to predicted failure.

Transformers

Power transformers on drill rigs are usually dry units, either convection or fan-cooled. When operated within their voltage, frequency, and load specification limits, these units provide years of continuous service. Heat damage resulting from overloads or insulation damage caused by environmental conditions are the most common causes of failure.

Planning the maintenance of power transformers results in longer life and reduced downtime. Reducing emergency or unexpected breakdowns on drill rigs means more time available for the drill functions. All power must be removed before performing maintenance functions.

Preventive Maintenance

1. Check for discoloration or damage resulting from overloads or moisture.
2. Keep transformer clean and free from dust, oil-laden material, and other foreign material.

3. Inspect connections for tightness.
4. Inspect insulation and test for voltage breakdown of winding insulation.
5. Inspect cooling blowers, if used, for proper operation and cooling.

Predictive Maintenance

1. If accurate maintenance records of insulation leakage currents are kept, it can be predicted to some degree when the transformer needs to be replaced before failure occurs.
2. Replace cooling blowers based on predicted life from vendors.

Cables and Connections

All electrical equipment depends on interconnecting cables to deliver power from one unit to the other and the terminations of these cables to complete the circuit. Integrity and reliability of the electrical system depends on the proper connection and the continued connection of the cables. Therefore, proper maintenance of the cabling system is very important.

On offshore rigs, the cable is usually protected in armor and further organized in a conduit or cable tray. Exposure to the environment is limited on these vessels, and cable reliability is usually high. Bolted connections must remain tight and plugs should be secure. Loose connections will appear through heated junctions, and increased voltage drops through these connections. Cable insulation exposed to caustic materials should always be cleaned and neutralized to prevent cable degradation.

Land rig cable systems are exposed to environmental conditions and to repeated handling and abuse during operations and rig moves. Inspection for insulation cuts, loose connectors, and loose terminations should be performed weekly during operations and after each rig-up. Power should be off prior to making maintenance inspections.

Preventive Maintenance

1. Inspect cables weekly for insulation cuts or nicks.
2. Check all AC and DC power cable terminations for tight connection; look for discoloration of bus bar or lugs indicating excessive heat generated by loose connection.

3. Clean and inspect all cables exposed to moisture, mud, or chemicals.

Predictive Maintenance

After analyzing the cable operating conditions, environment, and handling duty, determine the optimum time the cable and connectors should be replaced before a failure occurs. This may be several years or months, depending on its duty.

Electronic Equipment

With few exceptions, most electronic equipment is solid state and requires little maintenance. Keeping the equipment free from moisture, dust, fumes, and other foreign material or gases is essential for reliable operation. Periodic cleaning of the equipment, with power removed, ensures long- term operation.

Electric Brakes

The eddy current electric brake system used as an auxiliary brake on drawworks consists of four separate components: the eddy current brake, the power controller, the input transformer, and the driller's control.

The driller's control regulates braking power and is located near the driller during normal operation. It should be inspected regularly for proper operation, cleanliness, and proper cable terminations. Refer to the supplier's instructions for detailed information on the particular driller's control model used. A similar maintenance approach should be applied to the power controller and input transformer.

Maintainenance of the eddy current brake should adhere to the guidelines of the supplier. Since heat is generated in the brake rotor during the braking cycle, continuous rotation through the cooling water must be maintained. Maintain continuous coupling of the brake to the drawworks to ensure this action and long brake life. In addition, a continuous flow of cooling water at the recommended flow rate should be maintained without restriction. Finally, do not operate brake at excessive speeds for operator safety and wear on bearings.

Preventive Maintenance

1. Maintain proper lubrication to the bearings daily under normal drilling duty.
2. Inspect coil cavity drains and breathers for cleanliness and free access to air. Clean monthly.
3. Check air gap between rotor and magnet at inspection holes monthly for corrosion or scale due to poor quality water. Overhaul unit if required.
4. Inspect brake for proper alignment with drawworks as recommended by the manufacturer.
5. Check cooling water periodically for low mineral content; use corrosion-inhibiting additive, if necessary, as recommended by the brake manufacturer.
6. Check all interconnecting cables for solid connections and good insulation.
7. Check all brake coils for resistance continuity and isolation from brake housing (ground).
8. Brake voltage should be continuously variable from full off to full rating from controller. Check weekly.

Instruments

Many instruments are used throughout the drill rig, including generator control instruments and meters, SCR instruments, driller's console meters, drillstring loading instruments, and drill mud instruments. Each device provides measurements of various functions to help the rig manager, tool pusher, and driller perform drilling functions better.

The instruments include switchboard, panel, and portable functions and are usually driven by separate transducers. Switchboard-types are characterized by their high performance, ruggedness, and ease in reading, along with 1% or 1.5% accuracy. They are usually panel-mounted and come in sizes of 4½ in., 6 in., or 8¾ in.

Panel instruments cover a broader range, ranging in size from 1½ in. to 5½ in. cases, 0.5–5% accuracy, and wide price ranges. Portable instruments are usually the 0.5% taut-band types for both AC and DC measurements.

These instruments measure AC and DC voltage and current, real power, reactive power, power factor, temperature, frequency, phase, sync, speed, and other parameters.

Preventive Maintenance

1. Keep instruments clean.
2. Prevent the instruments from being exposed to shock and vibration.
3. Check for static charges causing errors in reading; use antistatic spray on meter faces if required.
4. Replace any broken meter faces.
5. Check meter accuracy with portable meters properly calibrated. Replace defective meters.
6. Check meter transducers such as current transformers (C.T.) and potential transformers (P.T.) for accuracy.

Establishing a Spare Parts Philosophy

After deciding to purchase new electric drill rig equipment, the next obvious decision is selecting spare parts to support rig operation. This should be weighed and supported by several factors:

- Is the equipment established in the drilling industry with parts stocked in supply stores, or is the equipment relatively new?
- Is the drillsite near a large city with adequate supply sources, or is it in a remote area where transportation or delivery is difficult?
- Will repairs be performed by a local rig crew, or are factory personnel required for repair?
- Does the drilling company employ several rigs in a concentrated area, or are they scattered throughout the country or offshore?
- Is the drilling operation located in one country with a source of supplies in another country, hampering export?

These and more specific questions must be evaluated and answered to the best interest of the drilling operation. In the event of equipment failure, parts are needed immediately for quick repair to return the equipment to drilling service. It is also apparent that spares of every drill rig item would be economically prohibitive. Therefore, an evaluation must be made to determine what spares must be kept at the rig, what spares must be kept at a central supply source, and what spares are less critical and can be ordered from the factory.

One spare parts philosophy is to categorize all rig items into levels based on criticality of the part. For example:

Level I:	Spares to be kept in sufficient quantity at the rig for immediate repair.
Level II:	Spares to be kept in a central company warehouse for one-to two-day shipment to the rig site.
Level III:	Spares could be ordered from the manufacturer within two to four weeks.

Using this philosophy of categorizing spare parts by level, analyze a typical system and prepare an abbreviated list of spare parts for the rig.

Rig manufacturer:	Houston, Texas
Customer location:	Mexico City, Mexico
Type rig:	12,000-ft land rig with 3 Cat 398 engines; 3 Kato generators; 4 GE 752 DC motors for mud pumps; drawworks and rotarytable; Baylor SCR; generator controls; and power distribution system.
Recommended spares (abbreviated):	
Level I:	1 engine actuator, proportional
(at rig)	1 Woodward electronic governor
	1 Basler SR8 voltage regulator
	1 reverse power relay
	1 set batteries for engine starting
	1 set SCR control printed circuit assemblies
	1 each, all SCR and control meters
	1 each, Martin-Decker load meters
	1 set SCR fuses
	1 Westinghouse SPB circuit breaker

	1 set power house plugs and receptacles
	1 set rig lights
Level II:	1 GE 752 DC motor (in Mexico)
	1 Kato generator
	1 Cat 398 diesel engine
	1 set mud pump liners
	1 reel drawworks cable
	1 each, mud mixing motors
Level III:	1 500-kva transformer(in U.S.A.)
	1 each, heaters for motor starters
	1 each, canvas wrap

Other factors that influence the spare parts philosophy are the reliability of each piece of equipment, support, and availability of manufacturer's service team, and capability of rig personnel to implement repairs.

Bibliography

"API Recommended Practice for Installation, Maintenance, and Operation of Internal Combustion Engines, API RP 7C-llF." Dallas: American Petroleum Institute, 1977.

"General Maintenance, Electrical Rotating Equipment, GEI-85 166." Erie, Pennsylvania: General Electric Co., 1977.

Westinghouse Electrical Maintenance Hints. Trafford, Pennsylvania: Westinghouse Electric Corp., 1974.

12 ELECTRIC EQUIPMENT–THEORY AND PERFORMANCE

Understanding the theory behind electric rig equipment operation removes the mystique associated with this equipment and allows rig personnel to operate and maintain it properly.

AC Motors

AC motors constitute a significant part of the equipment list on drill rigs since they are used in cooling blowers, mud-mixing systems, pumps, thrusters, and other applications. Speed is essentially constant since it depends primarily on the line frequency (usually 60 Hz) and secondarily on the motor's design characteristics.

A force is exerted when two magnetic fields developed by current-carrying conductors oppose each other. If the force is maintained as the conductors move away from each other, continuous motion can be achieved.

The North and South magnetic fields (N & S) create a magnetic field and resulting force from one pole to the other. If a current flows from the brushes and slip rings through the single conductor coil (rotor), an additional magnetic field develops around the conductor. These two magnetic fields interact, creating a force to rotate the coil.

In an induction AC motor, the stator field is an alternating field and the rotor consists of conductors located in slots. No external connection is made to the rotor circuit through such devices as slip rings or brushes.

As the alternating field builds up, goes to zero, then reverses each cycle, a current is "induced" in the rotor conductors, creating its own field. These two fields oppose each other, creating a force and resulting rotary motion. The rotor must maintain the speed of the alternating field of the stator, or its frequency. Any difference between the rotating field speed and the rotor speed is called slip speed.

Another factor determining rotor speed is the number of pairs of field windings placed on the stator. Each pair of windings is termed a pair of poles. In equation form, the synchronous rotor speed is:

$$N = \frac{120f}{P}, \text{ rpm}$$

where f = frequency of stator field, Hz
P = number of poles of field

A two-pole motor rotates at 3600 rpm when 60 Hz is applied to the field, and a four-pole rotates at 1800 rpm for 60 Hz. In actuality, all induction motors do not run at synchronous speed but run at a slight slip speed, resulting in rotor speeds of approximately 3450 rpm and 1750 rpm for two- and four-pole motors.

Three-phase motors are used predominantly on drill rigs because of their ability to develop torque at zero rotor speed without the aid of capacitors or auxiliary windings. These motors are simple and rugged and are considered the workhorse of the motor industry. They use three separate sets of field windings for each pair of poles connected to the three-phase supply and are distributed uniformly around the stator (Fig. 12–1). Since the three-phase voltages are phase-displaced by 120°, the resulting field is also phase-displaced by the same amount. The rotor also has three sets of windings for each pair of poles, and the induced field in the rotor creates a force resulting in rotary motion and torque.

"Squirrel cage" induction motors dominate the market and are so called because of the conductor bars pressed into the rotor structure. Each conductor bar on the rotor is terminated to a conductive end ring at both ends of the rotor. The ring's resistance determines the amount of induced current in the rotor and its torque-producing ability.

Reversing a three-phase motor is accomplished by reversing any two of the three-phase connections to the motor. This reverses the phase sequence of the applied field voltage and the direction of the rotating field on the stator.

Three-phase induction motors are classified by NEMA according to their torque speed characteristics. By changing the rotor resistance, several curves

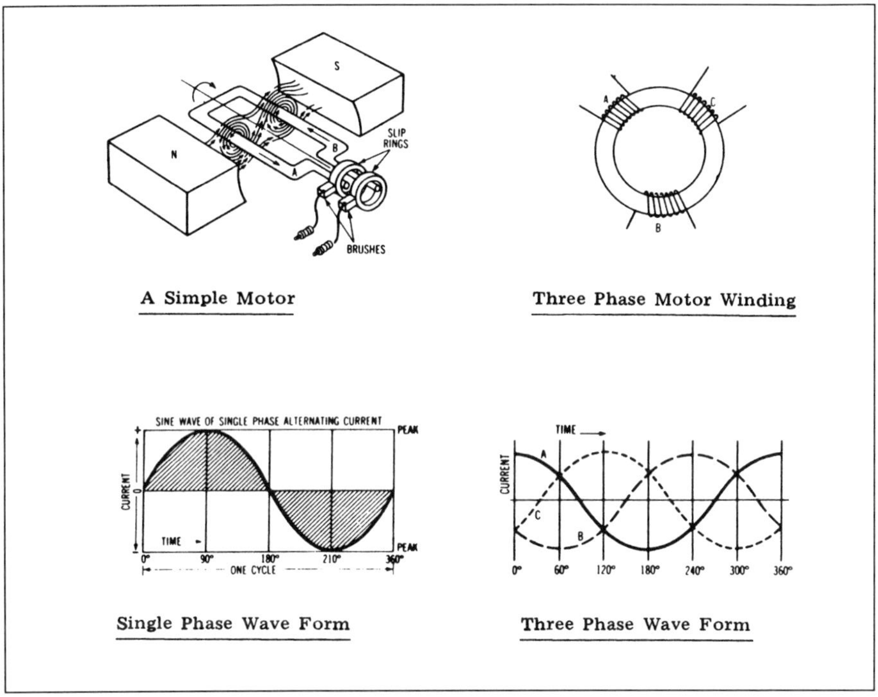

Fig. 12–1 Simple AC motor arrangement and waveforms

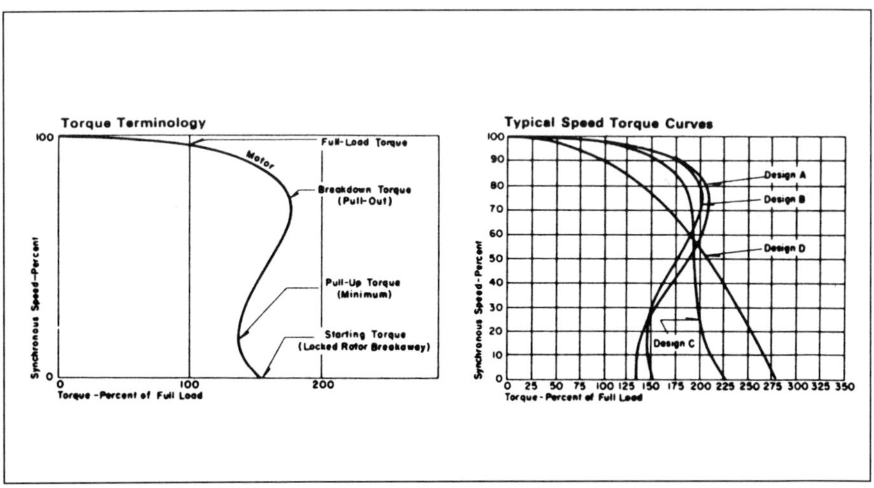

Fig. 12–2 Three phase induction motor characteristics

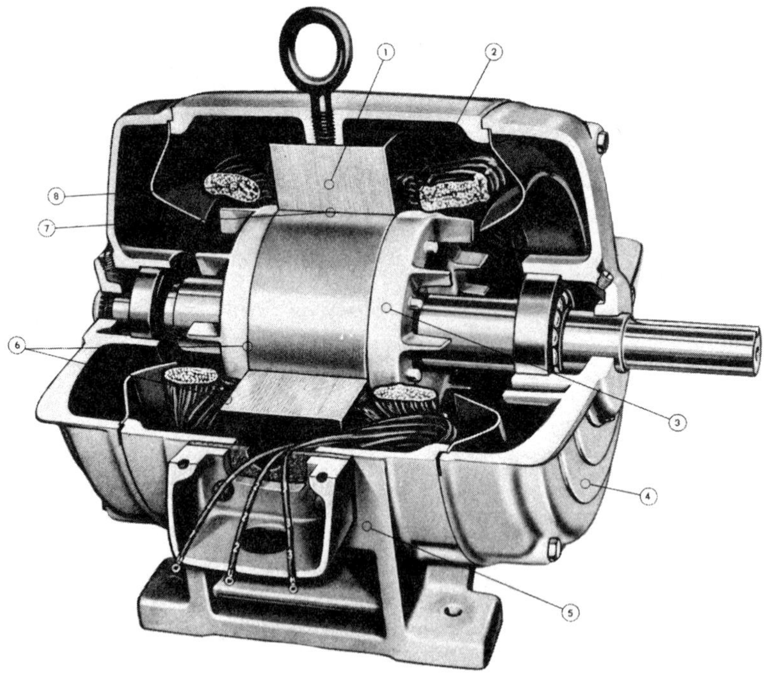

1. **Steel Laminated Core**
2. **Copper Stator Winding**
3. **Cast Aluminum Motor**
4. **Cast Iron Bearing Bracket**
5. **Cast Iron Housing**
6. **Stator and Rotor**
7. **Air Gap**
8. **Impregnated Winding**

Fig. 12–3 Typical three-phase AC induction motor construction (Courtesy Louis Allis Co.)

can be achieved; the basic characteristics, however, are defined as NEMA A, B, C, and D (Figs. 12–2, 12–3).

NEMA Design A or B (normal torque, low slip)

These designs have starting torques sufficient for accelerating a variety of drill rig loads. They are applied where slightly more than full load starting torque and low slip are required and where relatively high breakdown torque is needed to sustain occasional emergency overloads.

Design B starting currents are limited by NEMA, while Design A starting currents are not. Applications cover an extremely wide range, including drives for fans, pumps, drill presses, lathes, grinders, generators, certain types of machine tools, conveyors, and compressors.

NEMA Design C (high starting torque, low starting current, low slip)
Applications of this design are similar to those of Design B motors. The only difference is the starting torque must be 200% minimum. Applications for this design are machines in which the torque requirements are high at starting; they normally run at rated full load and are not subjected to high overload demands after running speed has been reached. Typical applications are electric stairways, pulverizers, compressors, and conveyors. In addition to proper rating and design requirements, care must be exercised in choosing the proper enclosure and meeting code requirements for motor applications or unusual environmental conditions.

NEMA Design D (high starting torque, low starting current, high slip)
Applications are high starting torque loads, such as oil pumping equipment. Four variations of AC motors enclosures are drip-proof, encapsulated, totally enclosed nonventilated or fan-cooled, and totally enclosed explosion-proof.

Drip-proof motors are installed in non-hazardous, reasonably clean surroundings, free of any abrasive or conducting dust and chemical fumes. Moderate amounts of moisture or dust and falling particles or liquids can be tolerated.

Encapsulated motors are installed in nonhazardous, high humidity or chemical applications free of clogging materials, metal dust, or chips and where hosing down or severe splashing is encountered.

Totally enclosed non-ventilated or fan-cooled motors are installed in non-hazardous atmospheres which contain abrasive or conducting dusts, high concentrations of chemical or oil vapors, and outdoor service.

Totally enclosed explosion-proof motors are installed in hazardous atmospheres containing:

- Class 1, Group D—acetone acrylonitrile, alcohol, ammonia, benzine, benzol, butane, ethylene dichloride, gasoline, hexane, lacquer solvent vapors, naphtha, natural gas, propane, propylene, styrene, vinyl acetate, vinyl chloride, or xylenes;

- •Class II, Group G—flour, starch, or grain dust
- •Class III, Group E—metal dust including magnesium and aluminum or their commercial alloys.

Under Class I only, there are two divisions which allow some latitude on motor selection. Generally, Class I, Division 1 locations are those in which the atmosphere is or may be hazardous under normal operating conditions. Included are locations which can become hazardous during normal maintenance. An explosion-proof motor is mandatory for Division 1 locations.

Class I, Division 2, refers to locations where the atmosphere may become hazardous only under abnormal or unusual conditions (breaking of a pipe, for example). In general, motors in standard enclosures can be installed in Division 2 locations if the motor has no normally sparking parts. Thus, open or standard totally enclosed squirrel cage motors are acceptable in some cases. Reference should always be made to applicable requirements for these applications. Motors with open slip rings or commutators (wound rotor, synchronous, or DC) are not allowed unless the commutators or slip rings are in an explosion-proof enclosure or purged with clean air. (See Appendix.)

Wound Rotor Induction Motor

This motor is constructed with three-phase stator windings and a wound rotor with three phases. Instead of a fixed rotor resistance, this motor uses brushes and slip rings to connect external resistors to the wound rotor for torque and speed control.

The motor operates in much the same way as the squirrel cage induction motor but has greater flexibility since its rotor resistance can be changed with variable resistance banks.

A wound rotor induction motor torque-speed characteristic can be varied (Fig. 12–4). These motors are useful in applications such as winches and hoists where variable use is intermittent.

DC Motors

The fundamental principle of operating AC motors can be applied to DC motors. Magnetic fields from the stator and rotor interact to develop force and rotary motion. Significant differences exist in DC motors because the field flux is stationary and not time-varying as in AC motors.

In order to develop a continuously rotating motor, maximum torque must be developed as each rotor conductor passes the peak field flux. This is accomplished by the rotor commutator, which automatically switches the current from one coil to another as it enters the maximum intensity of the field flux.

The commutator allows a DC motor to function. It consists of copper segments separated by mica insulators and terminated to rotor conductors on the machine. Power is delivered to the commutator by brushes sliding on its surface from a DC power source.

As the DC motor rotates, each rotor conductor moves through the stationary field flux and develops a voltage called counter emf. This counter emf opposes the applied voltage to the rotor coil and must be overcome by an additional amount of voltage to the coils. The relative direction of field flux to rotor coil flux determines the direction of rotation. Reversing the polarity of the field with respect to the rotor (or armature) causes reverse rotation.

Field excitation for a DC motor can be achieved in several different ways. The simplest technique is a series DC motor where the field and armature winding are electrically connected in series. Identical currents are present in the armature and field and can result in developing high torques at high current levels, since:

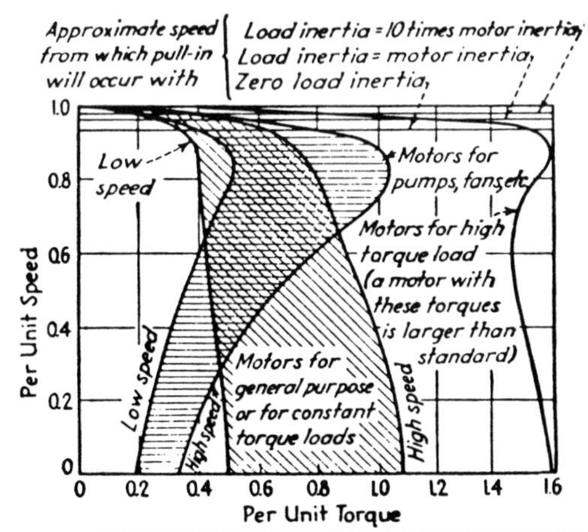

Fig. 12–4 Torque speed characteristics of wound-rotor induction motor

$$T_A = KI_AI_F \text{ ft-lbs (general equation)}$$

where T_A = DC motor torque, ft-lbs
K = machine constant
I_A = armature current, amps
I_F = field current, amps

However, a series motor can overspeed at light loads unless the motor voltage is reduced or removed (Fig. 12–5).

Shunt motors use a separately excited field from a fixed DC voltage source and develop a torque-speed curve as shown in Figure 12–6. Its field power requirements are small compared to the armature power and are typically 5%–6% of the armature power.

Motors which combine the features of series and shunt field motors are called compound motors.

Large DC machines have electromechanical efficiencies of more than 90% and provide variable speed with SCR or generator voltage controllers. Shunt motor speed can be varied by controlling the armature voltage with a fixed field.

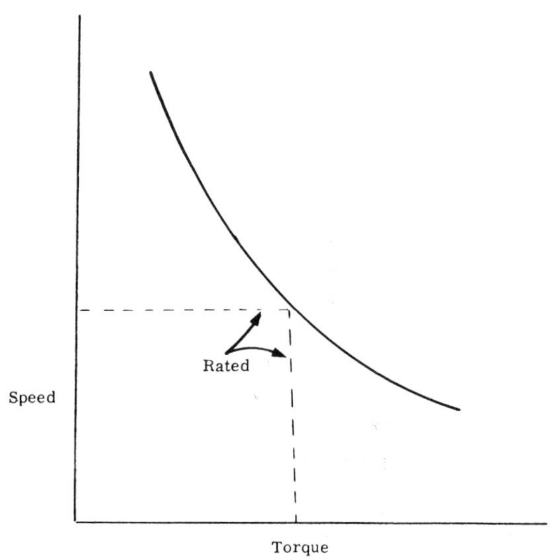

Fig. 12–5 Torque-speed characteristics of DC series motor at constant DC voltage

Some limited speed adjustment can also be achieved with field voltage changes. Series motor speed can be adjusted by varying the voltage and current to the motor, but speed stability is a function of the voltage and external load.

AC Generators

AC generators, or alternators, produce alternating current voltages proportional to their field excitation and speed. These units develop single-phase, three-phase, or polyphase voltages depending on the winding configuration at a frequency based on the following equation:

$$f = \frac{PN}{120}\ \text{Hz}$$

where f = frequency, Hz
P = number of machine poles
N = rotor speed, rpm

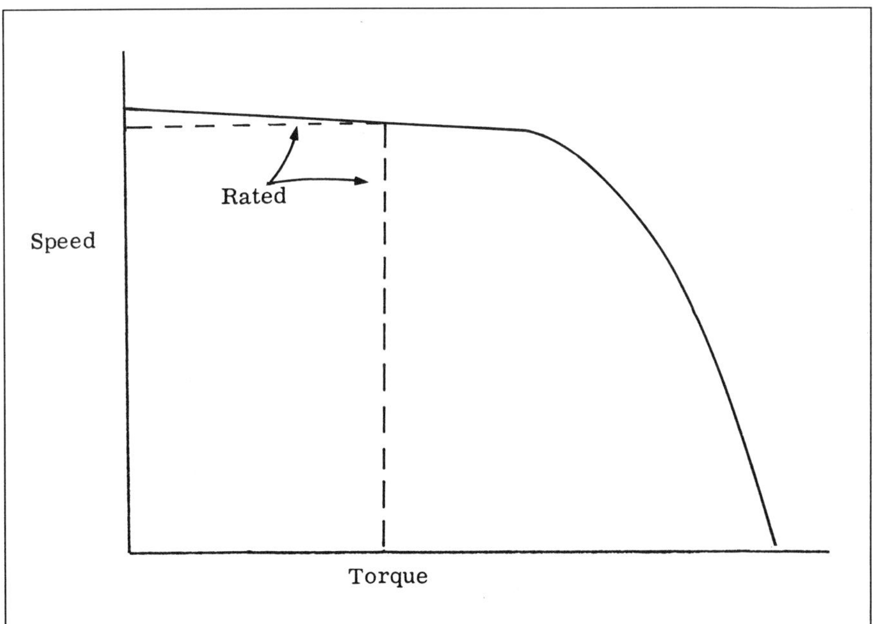

Fig. 12–6 Torque-speed characteristics of DC shunt motor at constant armature and field voltage

Three-phase generators usually consist of a separately excited DC field (usually on the rotor) and an armature composed of three-phase windings. As the rotor is turned by the prime mover, the field flux links with the armature conductors to generate or induce a voltage in the armature. At a particular speed, the generated voltage is proportional to the air gap flux density and a time-varying sinusoidal voltage is developed for each phase winding.

Brush-type generators use a commutator to transfer the current to the rotating field, while more modern machines use a separate AC generator to produce AC on its rotating field. This AC is then rectified with three-phase, full-wave rectifiers to DC for the field winding. These rectifiers rotate at the same speed as the field winding assembly thereby eliminating any brush/slip ring assembly (Fig. 12–7).

DC Generators

DC generators are becoming less common on oil drilling rigs, but a general understanding of their operation is necessary by those involved in drill rigs. The

Fig. 12–7 Cutaway view of AC generator (Courtesy General Electric Co.)

essential elements of a DC generator are shunt field, usually on the stator of the machine; armature, the power-generating portion of the machine; commutator, used to "rectify" and conduct the generated currents from the armature; and brushes, used to conduct the current from the commutator to the external circuit.

With the shunt field excited from a separate DC source, a prime mover turns the shaft of the generator. As the armature conductors pass through the magnetic field, a voltage is generated and conducted to the commutator segments. The magnitude of the voltage is proportional to both the magnetic field intensity and the speed of the armature. The equation that expresses this relationship is:

$$E = K\frac{d\varnothing}{dt}, \text{ volts}$$

where K = machine constant

$\frac{d\varnothing}{dt}$ = rate of change of flux

The polarity of the DC produced can be reversed by reversing the machine rotation or the field excitation current.

Transformers

A transformer is an electromagnetic device which changes the AC voltage or current in a circuit from one level to another. The term "transformer" means to transform voltage or current magnitude in a circuit at essentially the same kva level.

Transformers are constructed in several ways to suit particular applications; however, each one contains conductive coils wound around a metallic core or iron alloy. There are several types of transformers.

Isolation Transformer: One whose primary and secondary windings are electrically insulated from each other.

Autotransformer: A single winding transformer with a common terminal and tapped connections for voltage transforming; no electrical isolation is provided.

Signal Transformer: A small transformer for electrical signals, typically below power levels of 10 va.

Control Transformer: An intermediate-size transformer for control purposes usually rated 10–1,000 va.

Power Transformer: Used for power applications above 1 kva.

Single-phase Transformer: Transfers AC power in single-phase circuits.

Three-phase Transformer: Transfers AC power in three-phase circuits from one transformer assembly.

Potential Transformer: Usually transfers voltage at a relatively constant and precise level.

Current Transformer: Transfers current, usually at a high current level to a low current level; more secondary than primary turns.

Wye- or Delta-connected Transformer: In three-phase circuits, a wye connection terminates the three coils at a common point (neutral) and delivers power fronm the other three winding terminals; a delta connection terminates the three windings in a continuous connection resembling a triangle in its schematic and delivers power from each of the three connection points.

Variable Transformer: A single winding transformer resembling an autotransformer whose tap is a sliding carbon brush for variable voltage output.

Figure 12–8 shows an electrical schematic and construction of a single-phase transformer. The key elements in a transformer are the primary winding, secondary winding, and metallic core material. Wire size for each winding is a function of the current to be delivered in each one, and the core size is a function of the number of coil turns, voltage, and frequency applied.

Fundamental relationships between primary and secondary turns, voltage, and current are expressed in the following equation:

$$\frac{V_1}{V_2} = \frac{N_1}{N_2} = \frac{I_2}{I_1}$$

where V_1 = primary voltage
V_2 = secondary voltage
I_1 = primary current
I_2 = secondary current
N_1 = primary turns
N_2 = secondary turns

Note:
- Voltage varies directly with turns.
- Current varies inversely with turns.
- $\frac{N_1}{N_2}$ is the transformer "turns ratio."
- The product of voltage and current is a constant.
- $V_1I_1 = V_2I_2$ = constant kva.

The number of turns required for a winding are:

$$N = \frac{V}{K\text{Ø}f}, \text{ turns}$$

where V = voltage applied to coil
K = a design constant
Ø = magnetic flux in transformer core
f = frequency of voltage, Hz

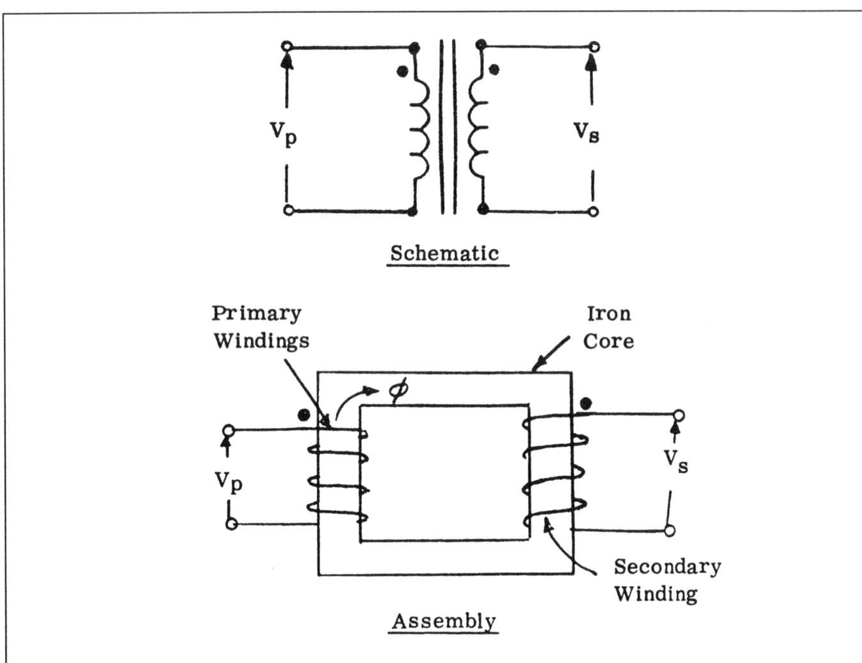

Fig. 12–8 Basic transformer diagrams

Note:
- Number of turns and voltage are proportional: The higher the voltage, the greater number of turns required.
- Number of turns is inversely proportional to the frequency: The higher the frequency, the fewer turns required.
- Number of turns is inversely proportional to the generated flux in the core: The better the core quality, the higher the flux allowed.

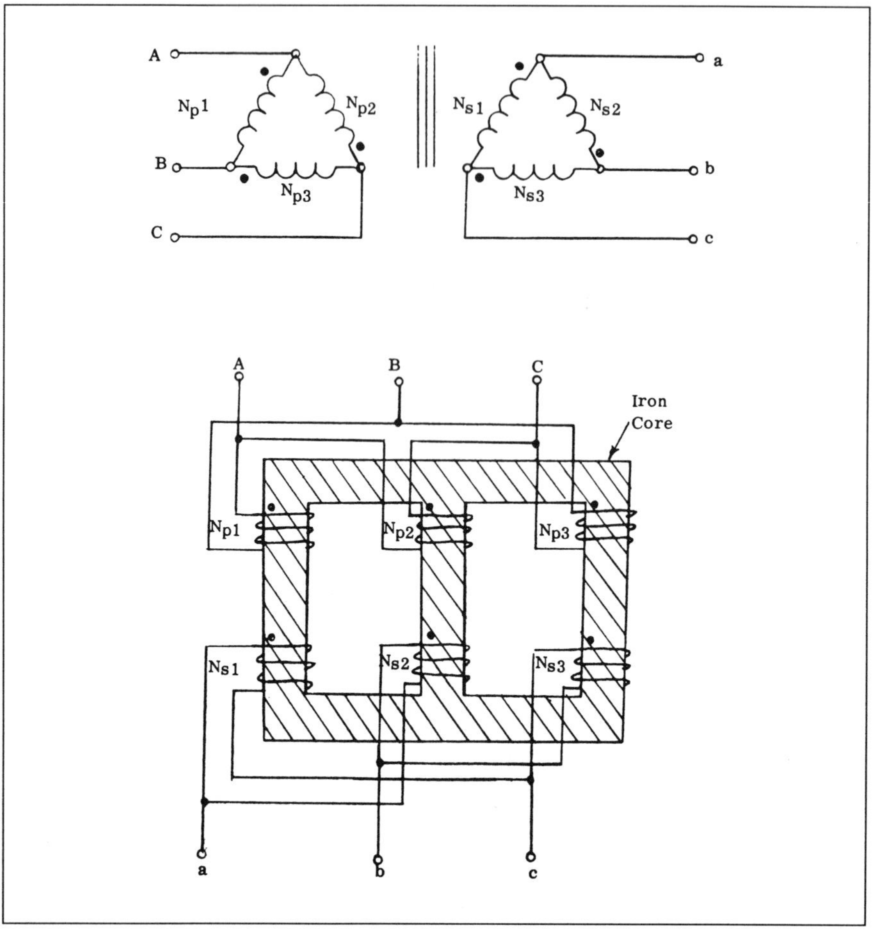

Fig. 12–9 Three-phase transformer schematic and assembly

The physical size of a transformer depends primarily on the volt-ampere product and the applied frequency. On most drill rig applications, the frequency is fixed at 60 Hz and the size is directly related to kva rating.

A three-phase transformer operates on a principle similar to the single-phase transformer except three coils are wound on three "legs" of the core. Since the flux in each coil is phase-displaced by 120°, the resultant magnetic flux in each leg is equal and displaced 120° apart (Fig. 12–9).

Other characteristics of transformers that are important to rig operation include:

Magnetizing Inrush Current—When voltage is first applied to a transformer, it may drive the core into magnetic saturations. When this occurs, the initial inrush current is limited only by the resistance and reactance of the primary winding until the core resets itself.

Exciting Current—The current that flows in the transformer primary with the secondary open-circuited.

Regulation—The voltage change on the secondary from no load to full load with constant primary voltage expressed in percent voltage droop or as percent impedance.

ANSI specifications for power transformers are defined in ANSI C57.12.80.

Bibliography

Fink, Donald M. *Standard Handbook for Electrical Engineers* New York: McGraw-Hill Book Co., 1968.

Westinghouse Electrical Maintenance Hints. Trafford, Pennsylvania: Westinghouse Electric Corp., 1974.

13 ELECTRICAL TECHNOLOGY ADVANCES

Electrical power on drilling rigs has allowed a significant number of advancements in control, power, and measuring equipment to assist in the drilling operation. Electrical technology allows cost reductions and improved performance through automation of drilling functions, continuous data logging of drilling activity, and greater control of power applications.

Downhole Drilling

Instead of using the rotary table to drive the drill bit from the surface, some systems use mud pressure to drive a hydraulic motor for powering the drill bit. The advantages of this sytem include directional drilling activity for exploratory and production drilling. A more recent application of this technology is horizontal drilling where multiple formations are penetrated for greater production.

Electrical instruments sense and guide the downhole tool under various drilling conditions, including programmed turns, bypassing a cement plug, returning to a sidetracked hole, and intersecting an existing hole to correct a blowout (Fig. 13–1).

Solar Power

Solar cells coupled with batteries are used increasingly as DC power supplies for critical instrumentation or controls on drill rigs. The solar cell panels are

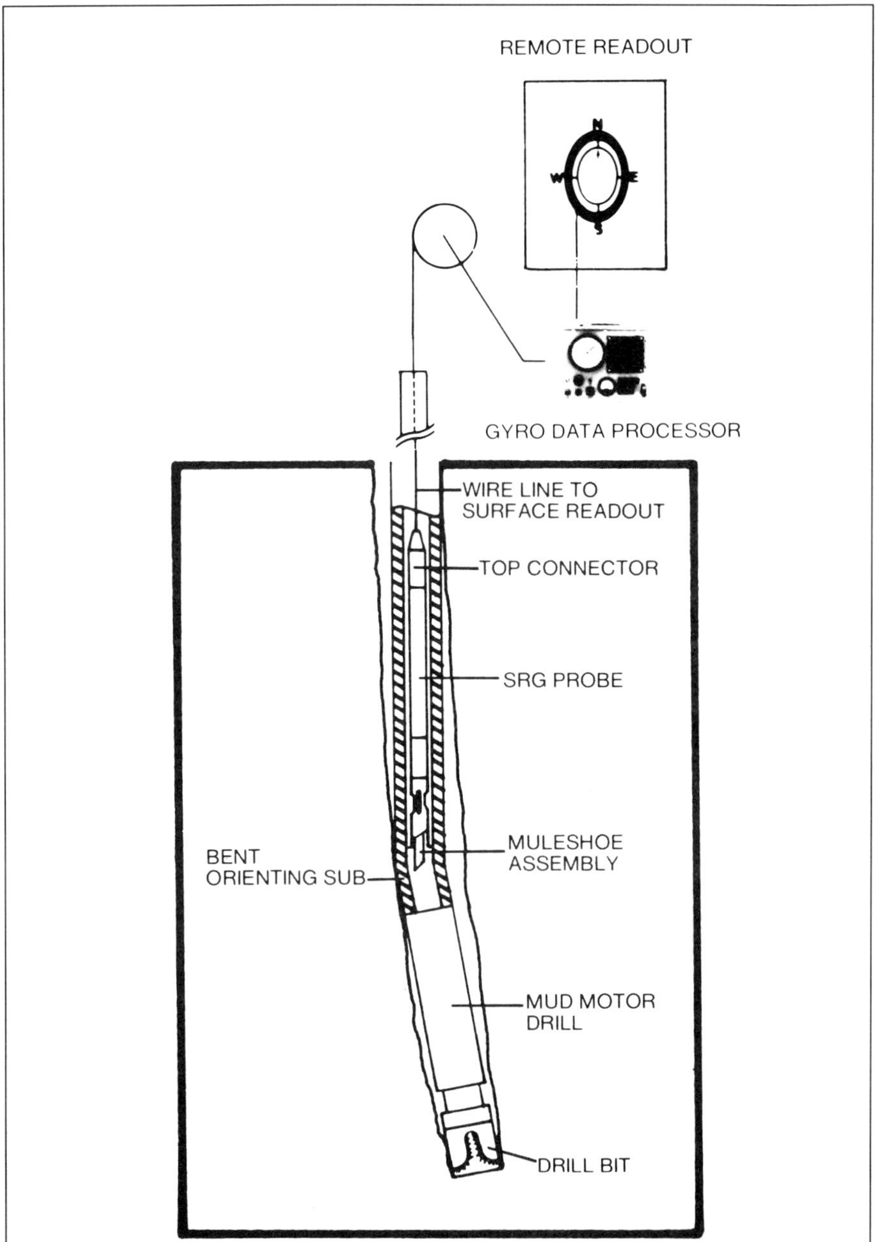

Fig. 13–1 Hole bypassing process

electrically connected and environmentally protected (Fig. 13–2).

The battery is the main source of DC power to the load, while the solar cell provides battery charging and some load current during sunlight periods. This system eliminates the need for electrical power from generators which may be important in the event of a power outage from the main power source.

Solar panels may be purchased in a variety of voltage and current ratings but

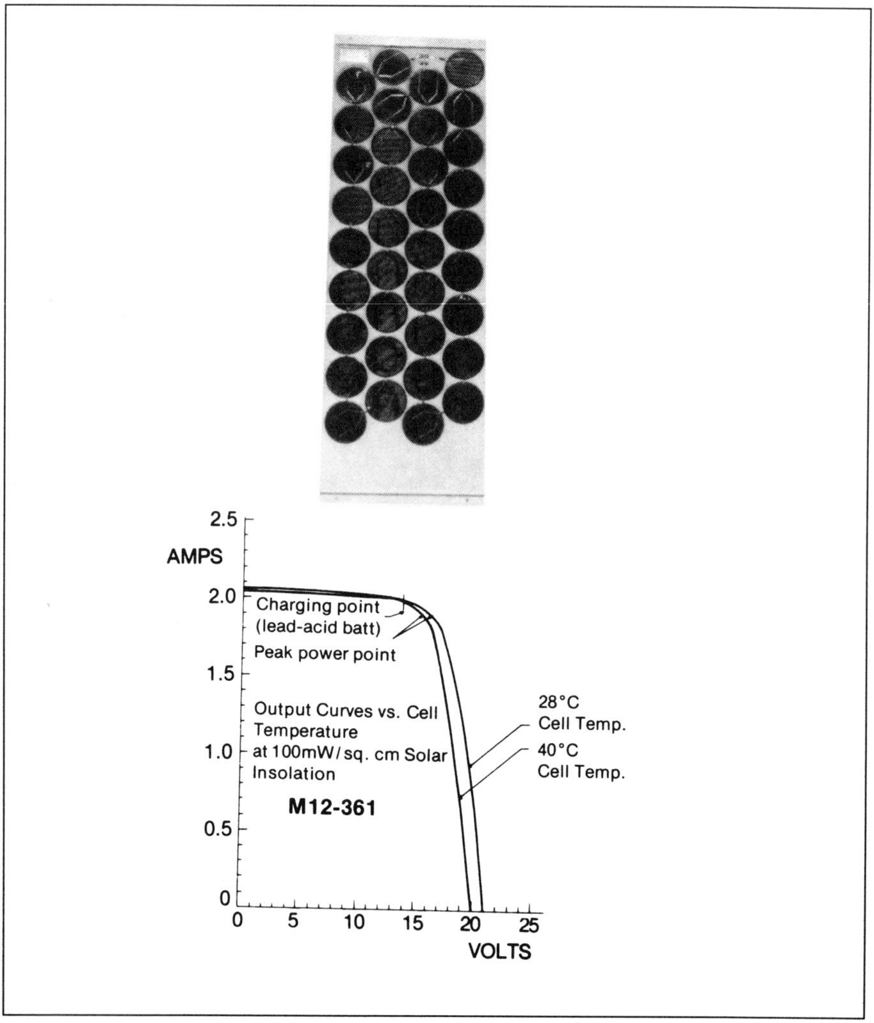

Fig. 13–2 Typical solar/battery DC supply schematic

the practical power obtained from a typical solar panel is 100 watts. This typical panel would come in increments of 10 watts and be connected into a system. (Fig. 13–3)

Low-Power SCR Systems

When SCR systems were first introduced, their initial advantages were in high-power drill systems up to 2500 HP. They were economical and flexible,

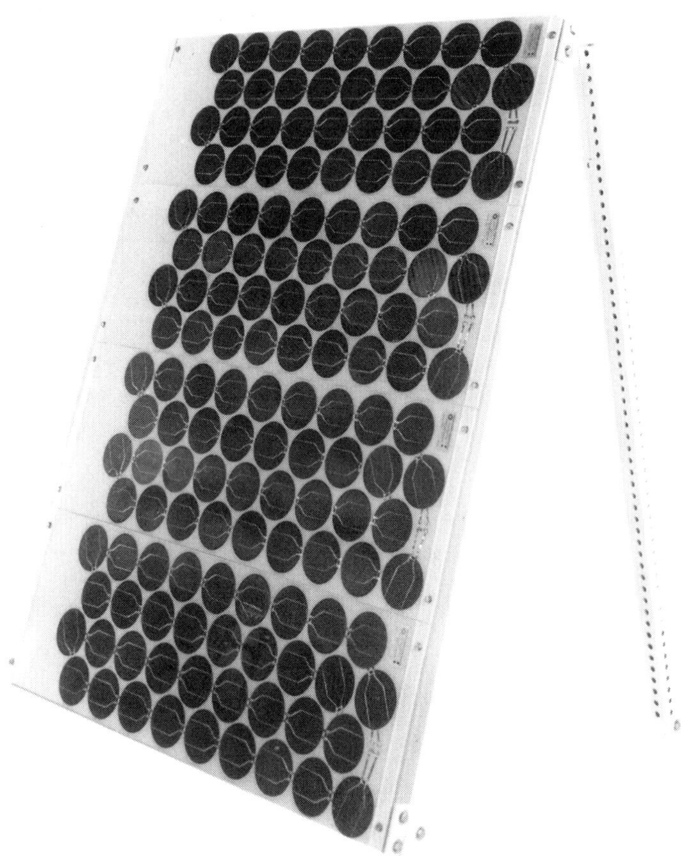

Fig. 13–3 Solar panel system (Courtesy Solar Power Corp.)

and they offered fuel savings advantages. As knowledge of SCR applications has improved, so has its use on smaller rigs.

Use of SCRs on small rigs whose drilling depth is less than 12,000 ft is on the increase as costs of the equipment drop and knowledge of its uses increases. Applications for well-service rigs and drill rigs in the 300–750 HP range are becoming more evident as energy needs increase and fuel resources become more scarce.

Rubber-tired rigs require that overland loads be as light as local conditions and laws will allow, and the small SCR system contributes to lighter weights. Ability to withstand vibrations and shock under hazardous road conditions without damage is essential during rig moves.

APPENDIX

Standards Agencies

IEEE Standards Documents[1]: The IEEE publishes several hundred standards documents covering various fields of electrical engineering.

National Electrical Manufacturers Association (NEMA) Standards[2]: The National Electrical Manufacturers Association prepares standards which establish dimensions, ratings, and performance requirements for electric equipment between manufacturers. Their standards are widely used in preparing purchase specifications.

National Fire Protection Association (NFPA) Standards Documents[3]: The National Fire Protection Association publishes standards documents specifying requirements for fire protection and safety. Of interest to industrial plant electrical engineers are the following:

- National Electrical Code (1990) NFPA No. 70, (ANSI C1-1990), (NEC)
- Handbook of the National Electrical Code, sponsored by the NFPA and published by McGraw-Hill. Contains the com-

---

[1]345 East 47 Street, New York. NY 10017
[2]155 East 44 Street, New York. NY 10017
[3]470 Atlantic Avenue, Boston, MA 02110

plete NEC text plus explanations edited to correspond with each edition of the NEC.

- Electrical Equipment Maintenance (1975) NFPA No. 70B (ANSI C132.1-1975)

Underwriters Laboratories Inc. (UL) Standards[4]: Underwriters Laboratories Inc. prepares safety standards for electric equipment including appliances and tests equipment for compliance with these standards. Manufacturers who have their products approved by UL may use the UL label on the equipment. UL periodically publishes lists of approved equipment.

The American National Standards Institute (ANSI)[5]: The ANSI does not write standards. It promotes and coordinates the development of American national standards and approves as American national standards those documents which have been prepared in accordance with ANSI regulations.

Standards which have been approved by other organizations, then approved as American National Standards, carry the identification numbers of both organizations and may be purchased from either. The sponsoring organization must keep the standard up to date. Standards carrying only an ANSI number were prepared by American National Standards committees organized and administered by other organizations in accordance with ANSI regulations. These committees are generally to coordinate participation by a large number of organizations.

Edison Electric Institute[6]: The Edison Electric Institute, the trade association of the privately owned electric utilities, publishes the following handbooks:

- *Electric Heating and Cooling Handbook*
- *A Planning Guide for Architects and Engineers*
- *Electric Space-Conditioning*
- *Industrial and Commercial Power Distribution*
- *Industrial and Commercial Lighting*

---

[4] 207 E. Ohio Street, Chicago. Il 60611
[5] 1430 Broadway, New York, NY 10018
[6] 90 Park Avenue, New York, N Y 10017

Occupational Safety and Health Administration (OSHA): Legislation by the U. S. government has had the effect of giving standards, such as those of the American National Standards Institute (ANSI), the impact of law. The Occupational Safety and Health Act, administered by the U. S. Department of Labor, permits federal enforcement of codes and standards. OSHA has adopted the 1990 National Electrical Code (NEC) for new electrical installations and equipment within the scope of subpart S-Electrical of OSHA regulations and also for major replacements, modifications, or repair installed after March 5, 1972. Some articles and sections of the NEC apply to all electrical installations and utilization equipment and thus are retroactive.

Environmental Studies: In all branches of engineering, social, ecological, and environmental concerns are being emphasized. Today's engineer must consider air, water, noise, and all other items which have an environmental impact. The limited availability of energy sources and the steadily increasing cost of electric energy require a concern with energy conservation on the part of the engineer.

The electrical engineer may participate in studies such as total energy compared to utility power, electric heating vs fossil fuel, a comparison of boilers, purchased steam, and the heat pump. The use of steam turbines compared to absorption units and electric drives for air conditioning may be evaluated. In these studies the effects of noise, vibration, exhaust gases, cooling methods, and energy requirements must be considered in relation to the immediate and sometimes general environment.

Handbooks:

The following handbooks are reputable in the electrical field. This list is not intended to be all- inclusive; other excellent references are available but are not listed here because of space limitations.

- *American Electricians' Handbook*, Crott, T., C. Carr, and J. Watt. New York: McGraw-Hill, 1970. The practical aspects of equipment, construction, and installation are covered.
- *Handbook*. American Society of Heating, Refrigerating, and Air-conditioning Engineers (ASHRAE).[7] This series of ref-

[7]345 East 47 Street, New York, NY 10017

erence books in four volumes, which are periodically updated, details the electrical and mechanical aspects of space-conditioning and refrigeration.

- *Electrical Transmission and Distribution Reference Book* Westinghouse Electric Corp., 1964.[8] The design and application of electric systems are outlined.
- *Electric Utility Engineering Reference Book.* Vol. 3. Westinghouse Electric Corp., 1965.[8] The application of high-voltage equipment and the design of high-voltage and network systems are covered in detail.
- *How to Design Electrical Systems.* McPartland, J., and the editors of Electrical Construction and Maintenance. McGraw-Hill, 1968. And Electrical Design Details. McPartland, J. and W. Novak. McGraw-Hill, 1960. These references offer an insight for the younger engineer into the systems approach to electrical design.
- *Industrial Power Systems Handbook.* Beeman, D. L., ed. McGraw-Hill, 1955. A text on electrical design with emphasis on equipment, including that applicable to commercial buildings.
- *Lighting Handbook.* Illuminating Engineering Society (IES),[9] 1972. All aspects of lighting, including seeing, recommended lighting levels, lighting calculations, and design, are included in extensive detail in this comprehensive text.
- *Standard Handbook for Electrical Engineers.* Fink, D. G. and J. M. Carroll, eds. McGraw-Hill, 1978. Virtually the entire field of electrical engineering is treated, including equipment and systems design.
- *Transformer Connections.* General Electric Co., publication GET-2.
- *Electrical Maintenance Hints.* Westinghouse Electric Corp, 1975.

[8]Printers' Division, Forbes Rd.. Trafford, PA 15085

[9]345 East 47 Street. New York, NY 10017

- *Inspection and Test of Electrical Equipment.* Shaw, E.T. Westinghouse Electric Corp., 1967.

Periodicals

Spectrum, the basic monthly publication of the IEEE, covers all aspects of electrical and electronic engineering. It contains references to IEEE books and other publications; technical meetings and conferences; IEEE group, society, and committee activities; abstracts of papers and publications of the IEEE and other organizations; and other material essential to the professional advancement of the electrical engineer. Following are some other well-known periodicals:

- *Actual Specifying Engineer,* 205 East 42 Street, New York, NY 10017
- *Electrical Consultant,* 1760 Peachtree Road NW, Atlanta, GA 30309
- *Lighting Design and Application,* Illuminating Engineering Society, 345 East 47 Street, New York, NY 10017
- *Plant Engineering,* 1301 South Grove Avenue, Barrington, IL 60010
- *Power,* 1221 Avenue of the Americas, New York, NY 10020
- *Power Engineering,* 1301 South Grove Avenue, Barrington, IL 60010

Manufacturers' Data

The electrical industry, through its associations and individual manufacturers of electrical equipment, issues many technical bulletins and data books. While some of this information is difficult for the individual to obtain, copies should be available to each major design unit.

The advertising sections of electrical magazines contain excellent material, usually well-illustrated and presented in a clear and readable form, concerning the construction and application of equipment. Such literature may be promotional; it may present the advertiser's equipment or methods in a "best light" and should be carefully evaluated. Manufacturers' catalogs are a valuable source of equipment information. Some of the larger manufacturers' complete catalogs are very extensive, covering dozens of volumes. However, these companies may issue abbreviated or condensed catalogs which are adequate for most applications.

Data sheets referring to specific items are almost always available from the sales offices. Some technical files may be kept on microfilm for use either by projection or by printing at larger design offices. Manufacturers' representatives, both sales and technical, can do much to provide complete information on a product.

Table A-1
Characteristics of Commonly Used Cable Insulation

Insulation	Maximum operating temperature of conductor	Applications environment	Range of AWG conductor size
RHW (Moisture- and heat-resistant rubber with moisture resistance and flame-retardant nonmetallic covering)	75°C(167°F)	Dry and wet locations	14–2000
THW (Moisture- and heat-resistant thermo-plastic with flame retardant)	75°C(167°F) or 90°C(194°F)	Dry and wet locations	14–2000
MTW (Moisture-, heat-, and oil-resistant thermoplastic with flame retardant)	60°C(140°F)	Machine tool wiring in wet locations	22–1000
	90°C(194°F)	Machine tool	22–1000 wiring in dry locations
TFE (Extruded polytetra-fluorsethylene)	250°C(482°F)	Dry location only	14–4/0
SIS (Synthetic, heat-resistant rubber)	90°C(194°F)	Switchboard wiring only	14–4/0

Table A-2
Allowable Current-Carrying Capacity in Amperes of Insulated Copper Conductors at 30°C Ambient

	Conductor temperature rating							
	60°C		*75°C*		*90°C*		*250°C*	
AWG Size	*In free air*	*In raceway*	*In free air*	*In raceway*	*In free air*	*In raceway*	*In free air*	*In raceway*
14	20	15	20	15	30	25	60	40
12	25	20	25	20	40	30	80	55
10	40	30	40	30	55	40	110	75
8	55	40	65	45	70	50	145	95
6	80	55	95	65	100	70	210	120
4	105	70	125	85	135	90	285	145
3	120	80	145	100	155	105	335	170
2	140	95	170	115	180	120	390	195
1	165	110	195	130	210	140	450	220
1/0	195	125	230	150	245	155	545	250
2/0	225	145	265	175	285	185	605	280
3/0	260	165	310	200	330	210	725	315
4/0	300	195	360	230	385	235	850	370
250	340	215	405	255	425	270	-	-
300	375	240	445	285	480	300	-	-
350	420	260	505	310	530	325	-	-
400	455	280	545	335	575	360	-	-
500	515	320	620	380	660	405	-	-
600	575	355	690	420	740	455	-	-
750	655	400	785	475	845	500	-	-
1000	780	455	935	545	1000	585	-	-
1500	980	520	1075	625	1260	700	-	-
2000	1055	560	1385	665	1470	775	-	-

*Not more than three conductors in raceway.

Table A-3

Correction Factors of Current-Carrying Capacities of Copper Conductors for Ambient Temperatures Over 30°C (86°F) Ambient

Ambient Temperature		Conductor Temperature Ratings			
°C	*°F*	*60°C*	*75°C*	*90°C*	*250°C*
40	104	0.82	0.88	0.91	-
45	113	0.71	0.82	0.87	-
50	122	0.58	0.75	0.82	-
55	131	0.41	0.67	0.76	-
60	140	-	0.58	0.71	0.95
70	158	-	0.36	0.58	0.91
75	167	-	-	0.50	0.89

Table A-4

Correction Factors of Current of Copper Conductors for More Than Three Conductors In A Raceway

Number of Conductors	Correction Factor
4 to 6	0.80
7 to 24	0.70
25 to 42	0.60
43 and more	0.50

Table A-5
Diameter and Area of Cables

	Insulation Type							
	RHW		THW		MTW		SIS	
AWG	Dia (in.)	Area (in.^{2})	Dia (in.)	Area (in.^{2})	Dia (in.)	Area (in.^{2})	Dia (in.)	Area (in.^{2})
14	0.204	0.0327	0.162	0.0206	0.131	0.4135	0.204	0.0327
12	0.221	0.0384	0.179	0.0251	0.148	0.0172	0.221	0.0384
10	0.242	0.0460	0.199	0.0311	0.163	0.0224	0.242	0.0460
8	0.328	0.0854	0.276	0.0598	0.245	0.0471	0.328	0.0854
6	0.397	0.1238	0.323	0.0819	0.323	0.0819	0.397	0.1238
4	0.452	0.1605	0.372	0.1087	0.372	0.1087	0.452	0.1605
3	0.461	0.1817	0.401	0.1263	0.401	0.1263	0.481	0.1817
2	0.513	0.2067	0.433	0.1473	0.433	0.1473	0.513	0.2067
1	0.568	0.2715	0.508	0.2027	0.508	0.2027	0.588	0.2715
1/0	0.629	0.3107	0.549	0.2367	0.549	0.2367	0.629	0.3107
2/0	0.675	0.3578	0.595	0.2781	0.595	0.2781	0.675	0.3578
3/0	0.727	0.4151	0.647	0.3288	0.647	0.3288	0.727	0.4151
4/0	0.785	0.4840	0.705	0.3904	0.705	0.3904	0.785	0.4840
250	0.868	0.5917	0.788	0.4877	0.788	0.4877	-	-
300	0.933	0.6837	0.843	0.5581	0.843	0.5581	-	-
350	0.985	0.7620	0.895	0.6291	0.895	0.6291	-	-
400	1.032	0.8365	0.942	0.6969	0.942	0.6969	-	-
500	1.119	0.9834	1.029	0.8316	1.029	0.8316	-	-
600	1.233	1.1940	1.142	1.0261	1.142	1.0261	-	-
700	1.339	1.4082	1.249	1.2252	1.249	1.2252	-	-
1000	1.494	1.7531	1.404	1.5482	1.404	1.5482	-	-
1500	1.801	2.5475	1.702	2.2748	1.702	2.2748	-	-
2000	2.021	3.2079	1.922	2.9013	1.922	2.9013	-	-

Table A-6
Color Code Per IEEE STD. 45-1977 Control and Signal Cables

Conductor number	Base color	Tracer color	Tracer color
1	Black		
2	White		
3	Red		
4	Green		
5	Orange		
6	Blue		
7	White	Black	
8	Red	Black	
9	Green	Black	
10	Orange	Black	
11	Blue	Black	
12	Black	White	
13	Red	White	
14	Green	White	
15	Blue	White	
16	Black	Red	
17	White	Red	
18	Orange	Red	
19	Blue	Red	
20	Red	Green	
21	Orange	Green	
22	Black	White	Red
23	White	Black	Red
24	Red	Black	White
25	Green	Black	White
26	Orange	Black	White
27	Blue	Black	White
28	Black	Red	Green
29	White	Red	Green
30	Red	Black	Green
31	Green	Black	Orange
32	Orange	Black	Green
33	Blue	White	Orange
34	Black	White	Orange
35	White	Red	Orange
36	Orange	White	Blue
37	White	Red	Blue
38	Brown		
39	Brown	Black	
40	Brown	White	

Conductor number	Base color	Tracer color	Tracer color
41	Brown	Red	
42	Brown	Green	
43	Brown	Orange	
44	Brown	Blue	

Table A-7
Table of Commonly Used Oilfield Constants and Equivalents

1 HP	=	746 w
1 HP	=	0.746 kw
π	=	3.1416
1 BTU	=	778.3 ft-lbs
1 circular mil	=	$\pi/4\ 10^{-6}$ sq in.
1 barrel	=	0. 15899 m^3
1 barrel	=	42 gal
1 cycle per second	=	1 Hz
1 mil	=	0.001 in.
Milli	=	10^{-3}
Micro	=	10^{-6}
1 mile	=	5280 ft
Nautical mile	=	1852 m
1 meter	=	3.2808 ft
1 ton	=	2000 lbs
1 inch	=	2.54 cm
1 cable	=	720 ft
1 cable	=	219.456 m
1 chain (engineer)	=	100 ft
1 chain (engineer)	=	30.48 m
1 fathom	=	6 ft
1 fathom	=	1.8288 m
1 acre	=	4,046.856 m^2
1 acre	=	43,560 ft^2
1 section	=	1 sq statute mi
1 township	=	36 sq statute mi
1 cubic meter	=	1000 liters
1 liter	=	61.024 $in.^3$
1 barrel	=	158.987 liters
1 cubic foot	=	28.3168 liters
1 gallon	=	3.7854 liters

Table A-7 (continued)Table of Commonly Used Oilfield Constants and Equivalents

1 knot	=	1.852 km/hr
1 knot	=	1.151 mph
1 inch of mercury, 32°F	=	13.6087 in.water @ 60°F
1 atmosphere	=	14.69596 lb/in.^{2}
1 watt-second	=	1 joule
1 joule	=	0.238846 cal
1 joule	=	9.478 x 10^4 BTUs
1 kilowatt hour	=	3,412.141 BTUs
1 horsepower	=	2.545.457 BTUs/hr
1 watt	=	3.412 BTUs/hr
1 footcandle	=	1.144 lumens/in.^{2}
1 BTU	=	0.0003929 HP hrs
Resistivity of standard annealed copper wire @25°C	=	10,575 ohm-cmil/ft
Temperature coefficient of resistance per °C of standard annealed copper wire (from 25°C)	=	0.00385/°C
Weight of standard annealed copper wire @2O°C	=	0.321 lbs
Coefficient of linear expansion per °C for standard annealed copper wire	=	0.0000169/°C

Table A-8
Temperature Conversion Chart

Celcius (°C)	**Farenheit (°F)**	**Absolute (°K)**
–273.15	–459.67	0.0
–200	–328	73.15
–180	–292	93.15
–160	–256	113.15
–140	–220	133.15
–120	–184	153.15
–100	–148	173.15

Celcius (°C)	Farenheit (°F)	Absolute (°K)
−80	−112	193.15
−60	−76	213.15
−40	−40	233.14
−20	− 4	253.15
−17.77	0	255.37
0	32	273.15
5	41	278.15
10	50	283.15
15	59	288.15
20	68	293.15
25	77	298.15
30	86	303.15
35	95	308.15
40	104	313.15
45	113	318.15
50	122	323.15
55	131	328.15
60	140	333.15
65	149	338.15
70	158	343.15
75	167	348.15
80	176	353.15
85	185	358.15
90	194	363.15
95	203	368.15
100	212	373.15
105	221	378.15
110	230	383.15
115	239	388.15
120	248	393.15
140	284	413.15
160	320	433.15
180	356	453.15
200	392	473.15
250	482	523.15
300	572	573.15
350	662	623.15
400	752	673.15
450	842	723.15
500	932	773.15

Table A-9
AC Motor Full Load Currents From National Electrical Code

	Three-phase AC induction motors squirrel cage & wound rotor (amps)			
hp	*115 VAC*	*220 VAC*	*460 VAC*	*575 VAC*
0.5	4.0	2.0	1.0	0.8
0.75	5.6	2.8	1.4	1.1
1	7.2	3.6	1.8	1.4
1.5	10.4	5.2	2.6	2.1
2	13.6	6.8	3.4	2.7
3	-	9.6	4.8	3.9
5	-	15.2	7.6	6.1
7.5	-	22.0	11.0	9.0
10	-	28.0	14.0	11.0
20	-	54.0	24.0	22.0
25	-	68.0	4.0	27.0
50	-	130.0	65.0	52.0
75	-	192.0	96.0	77.0
100	-	248.0	124.0	99.0
150	-	360.0	180.0	144.0
200	-	480.0	240.0	192.0

Table A-10
AC Motor Locked-Rotor Currents From National Electrical Code

	Three-phase AC induction motors squirrel cage & wound rotor (amps)			
HP	*115 VAC*	*230 VAC*	*460 VAC*	*575 VAC*
0.5 24	12.0	6.0	4.8	
0.75	36	16.8	8.4	6.8
1.0	42	21.0	10.8	8.4
1.5	60	30.0	15.0	12.0
2.0	78	39.0	19.8	15.6
3.0	-	54.0	27.0	24.0
5.0	-	90.0	45.0	36.0
7.5	-	132.0	66.0	54.0
10.0	-	162.0	84.0	66.0
15.0	-	240.0	120.0	96.0
20.0	-	312.0	156.0	126.0
25.0	-	384.0	192.0	156.0
50.0	-	750.0	378.0	300.0
75.0	-	1110.0	558.0	444.0
100.0	-	1475.0	738.0	588.0
150.0	-	2160.0	1080.0	864.0
200.0	-	2880.0	1440.0	1152.0

Table A-11
NEMA Starter Ratings

NEMA-Rated Full Voltage Motor Starters				
	Horsepower at three-phase voltages			Maximum current rating (Amps)
NEMA Size	230 VAC	460 VAC	575 VAC	
00	1.5	2	2	9
0	3	5	5	18
1	7.5	10	10	27
2	15	25	25	45
3	30	50	50	90
4	50	100	100	135
5	100	200	200	270
6	200	400	400	540

IEEE and ANSI Standards

IEEE and ANSI publish and promote concurrent standards for industry in designing, installing, operating, and maintaining electrical systems. Every five years, these standards are updated to replace all previous standards. For example, IEEE Std. 141-1976 is the standard for "Electric Power Distribution For Industrial Plants" and was published in 1976.

All standards may be obtained from the IEEE headquarters, 345 East 47 Street, New York, NY 10017. The following list represents those subjects published by IEEE which are suggested for the drilling industry.

IEEE Std	Subject
45	Recommended Practice for Electric Installations on Shipyards
66	Determination of Short-Circuit Characteristics of Direct Current Machinery
141	Recommended Practice for Electric Power Distribution for Industrial Plants
141	Standard Practices and Requirements for Thyristor Converters for Motor Drives
462	General Requirements for Distribution, Power, and Regulating Transformers
508	Low-Voltage AC Power Circuit Protectors Used in Enclosures
94	Definition of Terms for Automatic Generator Control of Electric Power Systems
91	Graphic Symbols for Logic Diagrams (Two-State Devices) (ANSI Y32.14-1973)
100	IEEE Standard Dictionary of Electrical and Electronics Terms (ANSI C42.100- 1972)
260	Letter Symbols for Units Used in Science and Technology (ANSI Y10.19-1969)
268	Units in Published Scientific and Technical Work
280	Letter Symbols for Quantities Used in Electricai Science and Electrical Engineering (ANSI YIO.5-1968)
315	Graphic Symbols for Electrical and Electronics Diagrams (ANSI Y32.2-1975) (CSA Z99-1975)

142	Grounding of Industrial and Commercial Power Systems (IEEE Gray Book) (ANSI C114.1-1973)
241	Electric Power Systems in CommercialBuildings (IEEE Gray Book)
242	Protection and Coordination of Industrial and Commercial Power Systems (IEEE Buff Book)
446	Emergency and Standby Power Systems(IEEE Orange Book)
255	Letter Symbols for Semiconductor Devices

National Electrical Code References

The following are applicable 1990 NEC sections that are useful in the design of electric rigs:

Article	Description
110	Electrical installation
200	Wiring design and protection
210	Branch circuit wiring
215	Feeder wiring
220	Branch circuit and feeder calculation
230	Service wiring
240	Overcurrent protection
250	Grounding and bonding practices
280	Lightning arrestors
300	Wiring installation methods
305	Temporary wiring methods
310	Conductor types and characteristics
318	Continuous rigid cable supports
330–339	Cable types
346	Rigid metal conduit
347	Rigid nonmetallic conduit
348	Electrical metallic tubing
350	Flexible metal conduit
351	Liquid-tight flexible metal conduit
352	Surface raceways

353	Multi-outlet assemblies
354–358	Raceways
362	Wireways
364	Busways
370	Outlet, switch and junction boxes
373	Cabinets and cutout boxes
380	Switches
384	Switchboards and panelboards
390	Prefabricated buildings
400	Flexible cords and cables
402	Fixture wires
410	Lighting fixtures, receptacles
424	Fixed electric space heaters
430	Motors, motor controls
450	Transformers
460	Capacitors
470	Resistors, reactors
480	Batteries
500	Hazardous locations
610	Cranes and hoists
630	Electric welders
700	Emergency systems
710	Over 600 v
720	Under 50 v
725	Remote control, signal circuits
730	Outside branch circuits and feeders

NEC Definitions and Classifications of Hazardous Areas*

Class I Locations

Class I locations may have flammable gases or vapors present in the air in quantities sufficient to produce explosive or ignitible mixtures. Class 1 locations shall include the following:

Class 1, Division 1

Locations in which (1) hazardous concentrations of flammable gases or vapors exist continuously, intermittently, or periodically under normal operating conditions, (2) hazardous concentrations of such gases or vapors may exist frequently because of repair or maintenance operations or because of leakage, or (3) breakdown or faulty operation of equipment or processes which might release hazardous concentrations of flammable gases or vapors might also cause simultaneous failure of electrical equipment.

(This classification usually includes locations where volatile flammable liquids or liquefied flammable gases are transferred from one container to another; interiors of spray booths and areas in the vicinity of spraying and painting operations where volatile flammable solvents are used; locations containing open tanks or vats of volatile flammable liquids; drying rooms or compartments for the evaporation of flammable solvents; locations containing fat and oil extraction apparatus using volatile flammable solvents; portions of cleaning and dyeing plants where hazardous liquids are used; gas generator rooms and other portions of gas manufacturing plants where flammable gas may escape; inadequately ventilated pump rooms for flammable gas or for volatile flammable liquids; the interiors of refrigerators and freezers in which volatile, flammable materials are stored in open, lightly stoppered, or easily ruptured containers, and all other locations where hazardous concentrations of flammable vapors or gases are likely to occur in the course of normal operations.)

Class 1, Division 2

Locations (1) in which volatile flammable liquids or flammable gases are handled, processed, or used, but in which the hazardous liquids, vapors, or gases will normally be confined within closed containers or closed systems from which they can escape only in case of accidental rupture or breakdown of such containers or systems, or in case of abnormal operation of equipment, (2) in which hazardous concentrations of gases or vapors are normally prevented by positive mechanical ventilation, but which might become hazardous through failure or abnormal operation of the ventilating equipment, or (3) which are adjacent to Class I, Division I locations, and to which hazardous concentra-

---

tions of gases or vapors might occasionally be communicated unless such communication is prevented by adequate positive-pressure ventilation from a source of clean air and effective safeguards against ventilation failure are provided.

(This classification usually includes locations where volatile flammable liquids or flammable gases or vapors are used, but which, in the judgment of the authority having jurisdiction, would become hazardous only in case of an accident or of some unusual operating condition. The quantity of hazardous material that might escape in case of accident, the adequacy of ventilating equipment, the total area involved, and the record of the industry or business with respect to explosions or fires are factors that should receive consideration in determining the classification and extent of each hazardous area.

Piping without valves, checks, meters, and similar devices would not ordinarily be deemed to introduce a hazardous condition even though used for hazardous liquids or gases. Locations used for the storage of hazardous liquids or of liquefied or compressed gases in sealed containers would not normally be considered hazardous unless subject to other hazardous conditions also.

Electrical conduits and their associated enclosures separated from process fluids by a single seal or barrier shall be classed as Division 2 locations if the outside of conduit and enclosures is a nonhazardous area.)

Class II Locations

Class II locations are those which are hazardous because of the presence of combustible dust. Class II locations shall include the following:

Class II, Division 1

Locations (1) in which combustible dust is or may be in suspension in the air continuously, intermittently, or periodically under normal operating conditions, in quantities sufficient to produce explosive or ignitible mixtures, (2) where mechanical failure or abnormal operation of machinery or equipment might cause such mixtures to be produced and might also provide a source of ignition through simultaneous failure of electrical equipment, operation of protection devices, or from other causes, or (3) in which dusts of an electrically conducting nature may be present.

(This classification usually includes the working areas of grain handling and storage plants; rooms containing grinders or pulverizers, cleaners, graders, scalpers, open conveyors or spouts, open bins or hoppers, mixers or blenders, automatic or hopper scales, packing machinery, elevator heads and boots, stock distributors, dust and stock collectors (except all-metal collectors vented to the outside), and all similar dust-producing machiners and equipment in grain processing plants, starch plants, sugar pulverizing plants, malting plants, hay grinding plants, and other occupancies of similar nature; coal pulverizing plants (except where the pulverizing equipment is essentially dust-tight); all working areas where metal dusts and powders are produced, processed, handled, packed, or stored (except in tight containers); and all similar locations where combustible dust may, under normal operating conditions, be present in the air in quantities sufficient to produce explosive or ignitible mixtures.

Combustible dusts which are electrically nonconducting include dusts produced in the handling and processing of grain and grain products, pulverized sugar and cocoa, dried egg and milk powders, pulverized spices, starch and pastes, potato and woodflour, oil meal from beans and seed, dried hay, and other organic materials which may produce combustible dusts when processed or handled. Electrically conducting nonmetallic dusts include dusts from pulverized coal, coke, and charcoal. Dusts containing magnesium or aluminum are particularly hazardous and every precaution must be taken to avoid ignition and explosion.

Class II, Division 2

Locations in which combustible dust will not normally be in suspension in the air or will not be likely to be thrown into suspension by the normal operation of equipment or apparatus, in quantities sufficient to produce explosive or ignitible mixtures, but (1) where deposits or accumulations of such dust may be sufficient to interfere with the safe dissipation of heat from electrical equipment or apparatus, or (2) where such deposits or accumulations of dust on, in, or in the vicinity of electrical equipment might be ignited by arcs, sparks, or burning material from such equipment.

(Locations where dangerous concentrations of suspended dust would not be likely, but where dust accumulations might form on or in the vicinity of electrical equipment would include rooms and areas containing only closed spouting and conveyors, closed bins or hoppers, or machines and equipment from which appreciable quantities of dust would escape only under abnormal operating conditions; rooms or areas adjacent to locations described in Division 1, and into which explosive or ignitible concentrations of suspended dust might be communicated only under abnormal operating conditions; rooms or areas where the formation of explosive or ignitible concentrations of suspended dust is prevented by the operation of effective dust control equipment; warehouses and shipping rooms where dust-producing materials are stored or handled only in bags or containers; and other similar locations.)

Class III Locations

Class III locations are hazardous because of the presence of easily ignitible fibers or flyings, but in which such fibers or flyings are not likely to be in suspension in air in quantities sufficient to produce ignitible mixtures. Class III locations shall include the following:

Class III, Division 1

Locations in which easily ignitible fibers or materials producing combustible flyings are handled, manufactured or used. Such locations usually include some parts of rayon, cotton and other textile mills; combustible fiber manufacturing and processing plants; cotton gins and cotton-seed mills; flax processing plants; clothing manufacturing plants; woodworking plants; and establishments and industries involving similar hazardous processes or conditions.

Easily ignitible fibers and flyings include rayon, cotton (including cotton linters and cotton waste), sisal or henequen, istle, jute, hemp, tow, cocoa fiber, oakum, baled waste kapok, Spanish moss, excelsior, and other materials of similar nature.

Class III, Division 2

Locations in which easily ignitible fibers are stored or handled (except in process of manufacture).

Electrical Equations

1. Ohm's Law: $V = IR$

R = resistance, ohms
V = voltage, volts
I = current, amperes

or, $V = IZ$

Z = impedance, ohms
V = voltage, volts
I = current, amperes

2. $/Z/ = \sqrt{R^2 + X^2}$

/Z/	=	absolute value of impedance, ohms
R	=	resistance, ohms
X	=	reactance, ohms

3. $\omega = 27\pi f$ radians/sec

ω	=	angular velocity, rad/sec
f	=	frequency, Hz

4. $X_L = 2\pi fL = \omega$ ohms

X_L	=	inductive reactance, ohms
f	=	frequency, Hz
L	=	inductance, henries
ω	=	angular velocity, rad/sec

5. $Xc = \dfrac{1}{2\pi fc} = \dfrac{1}{C}$ ohms

Xc	=	capacitive reactance, ohms
f	=	frequency, Hz
C	=	capacitance, farads
	=	angular velocity, rad/sec

6. $P = I^2R = \dfrac{V^2}{C}$ watts

P	=	real power, watts
I	=	RMS current, amperes
R	=	resistance, ohms
V	=	RMS voltage, volts

7. $P = VI(\cos\theta) = VI(F_p)$, watts

P	=	real power, watts
V	=	RMS voltage, volts
I	=	RMS current, amperes
$\text{Cos}\,\theta$	=	Fp = power factor
θ	=	angle between current and voltage

8. $P = kva\,(F_p)\ kw$

P = real power, kw
kva = kilovoltamperes
F_p = power factor

9. $kva_3 = \dfrac{\sqrt{3}\,V_L I_L}{1000}$

kva_3 = apparent power of three phase, kva
V_L = Line-line voltage, volts
I_L = Line current, amperes

10. $kva = \dfrac{VI}{1000}$ kilovolt-amperes

kva = apparent power of single phase
V = single-phase voltage
I = line current, amperes

11. $\text{Efficiency} = \dfrac{\text{power out}}{\text{power in}} \times 100\%$

12. $HP = \dfrac{TN}{5252} \quad HP = \dfrac{FV}{33{,}000}$

HP = horsepower
T = torque in ft-lbs
N = speed in rpm
V = linear velocity in ft/min
F = force in lbs

13. $F_p = \dfrac{kw}{kva}$, power factor

F_p = power factor
kw = kilowatts
kva = kilovoltamperes

14. $F_p = \sin\theta = \dfrac{\text{kvars}}{\text{kva}}$

F_p = reactive factor
Sin θ = sine of angle between current and voltage
kvars = reactive power
kva = apparent power

15. W_h = P x t, watt-hrs

W_h = energy, watt-hrs
P = power, watts
t = elapsed time, hrs

16. $R = \dfrac{I}{G}$

R = resistance, ohms
G = conductance, mhos
I = current, amperes

17. $R = \rho\dfrac{l}{A}$, ohms

R = resistance of electrical conductor, ohms
ρ = resistivity of conductor material,ohm-cmil/ft
l = length of conductor, ft
A = area of conductor, circular mils
(ρ = 10.575 ohm cmil/ft for standard annealed copper wire @ 25°C.)

18. $C = \dfrac{Q}{V}$ farads

C = capacitance, farads
V = voltage, volts
Q = electrical charge, coulombs

19. $L = \dfrac{e}{di/dt}$ henries

L = inductance, henries
e = induced voltage, volts
di/dt = rate of change of current, amps/sec

20. $I = dQ/dt$ amperes

I = amperes
dQ/dt = rate of change of electrical charge, coulombs/sec

21. $\frac{N_p}{N_s} = \frac{V_p}{V_s} = \frac{I_s}{I_p}$ (turns ratio)

In a transformer:
N_p = primary turns
N_s = seconday turns
V_p = primary voltage, volts
V_s = secondary voltage, volts
I_s = secondary current, amperes
I_p = primary current, amperes

22. $°C = 5/9\ (°F - 32)$

23. $°F = 9/5\ (°C + 32)$

24. $°K = °C + 273.15$

25. $°R = °K/1.8$

26. $f = \frac{PN}{120}$ Hz

f = frequency of AC generator, Hz
P = number of poles of AC generator
N = speed of armature of AC generator, rpm

27. $N = \frac{120f}{P}$, rpm

N = synchronous speed of AC motor, rpm
f = supply frequency, Hz
P = number of motor poles

Bibliography

Fink, Donald G. *Standard Handbook for Electrical Engineers,* New York, McGraw-Hill Co., 1978.
National Electrical Code, Boston: National Fire Protection Association, 1990.

GLOSSARY

AWG	Commonly used gage for electrical wires in America is the American Wire Gauge.
Active power	Time rate of flow of electrical energy; real component of apparent power; in watts, symbol is w or kw; power symbol is P.
Ampere turns	The product of current and turns in a coil of wire; symbol is NI.
Apparent power	The product of rms voltage and current; in volt-amperes; symbol is va or kva.
Armature	The power-handling portion of a DC machine.
Battery	An electrochemical device whose chemical reaction with its active elements converts chemical energy to DC electrical energy.
Brushes	Usually constructed from carbon, these devices slide on the commutator of a DC machine to carry the rotor current to/from the external circuit.
Bus bar	A solid, metallic conductor rigidly supported to carry heavy currents.

Capacitance	Electrical factors in a circuit that store electrical charge and whose impedance is inversely proportional to the rate of change of voltage across it in farads; symbol is C.
Commutator	Consists of hard copper bars with insulators between them to rectify and conduct current to and from therotor of a DC machine.
Conductance	The factor in an electrical circuit that allows current flow; reciprocal of resistance in ohms; symbol is G.
Current	The transfer of electrons per unit time in an electrical conductor; in amperes; symbol is I.
Field	The excitation portion of a DC machine.
Flux density	The concentration of magnetic flux in a given area in lines per square inch; symbol is B.
Frequency	Cyclic (alternating) electrical quantity variation in cycles per second or Hertz; symbol is Hz.
Horsepower	An analogy of the unit of power, the watt; symbol is HP.
Impedance	The total resistance to current flow in an AC circuit representing the vectoral sum of resistance, capacitive reactance and inductive reactance in ohms; symbol is Z.
Inductance	Electrical factors in a circuit that resist changes in current and whose impedance is directly proportional to the rate of change of voltage across it in henries; symbol is L.
Insulator	Material placed between conductors to prevent electrical conduction.
Magnetic Flux	The measurement of magnetic force produced by a magnetic field in Lines or Webers; symbol is Ø.
Phase Angle	Electrical angle difference in degrees (or radians) between current and voltage in the same circuit; symbol is θ.

Phases	The number of electrical terminals displaced in time phase in an AC circuit.
Power factor	Ratio of active power to apparent power nondimensional quantity; symbol is Fp.
Reactance	The AC impedance of a capacitor or inductor; the imaginary part of impedance in ohms; symbol is X
Reactive factor	Ratio of reactive power to apparent power non-dimensional quantity; symbol is Fq.
Reactive power	The quadrature component of apparent power in vars; symbol is Pq.
Resistance	Electrical dissipative factors in a circuit that limit the flow of current from a voltage source in ohms; symbol is R.
Skin effect	The effect produced by large-diameter conductors or high frequencies where the current density is greater at the surface than the inner part of the conductor; net result is an increase in effective resistance.
Slip speed	Represents the difference between rotor or shaft rpm and the rpm of the rotating magnetic field in the stator (synchronous speed) and is normally expressed as a percent of synchronous speed.
Slip rings	Sliding surface contacts to conduct half-cycle phase power from the rotor of AC machines.
Subtransient reactance	$X''d \frac{1}{M}$ The equivalent inductive stator obtained from the direct axis of an AC generator; it is the significant short term (under six cycles) reactance considered when evaluating fault currents delivered by AC generators.
Transformer	An electromagnetic device used to convert voltages from one level to another and provide voltage isolation.
Turns ratio	The ratio of the primary turns to secondary turns of a

	transformer which represents the ratio of primary voltage to secondary voltage of a transformer.
Voltage	Potential difference between two terminals or points in an electrical circuit in volts; symbol is V.

INDEX

A

B

C

D

E

F

G

H

I

J

K

L

M

N

O

P

R

S

T

U

V

W